National Aeronautics and Space Administration

Adventures in Rocket Science

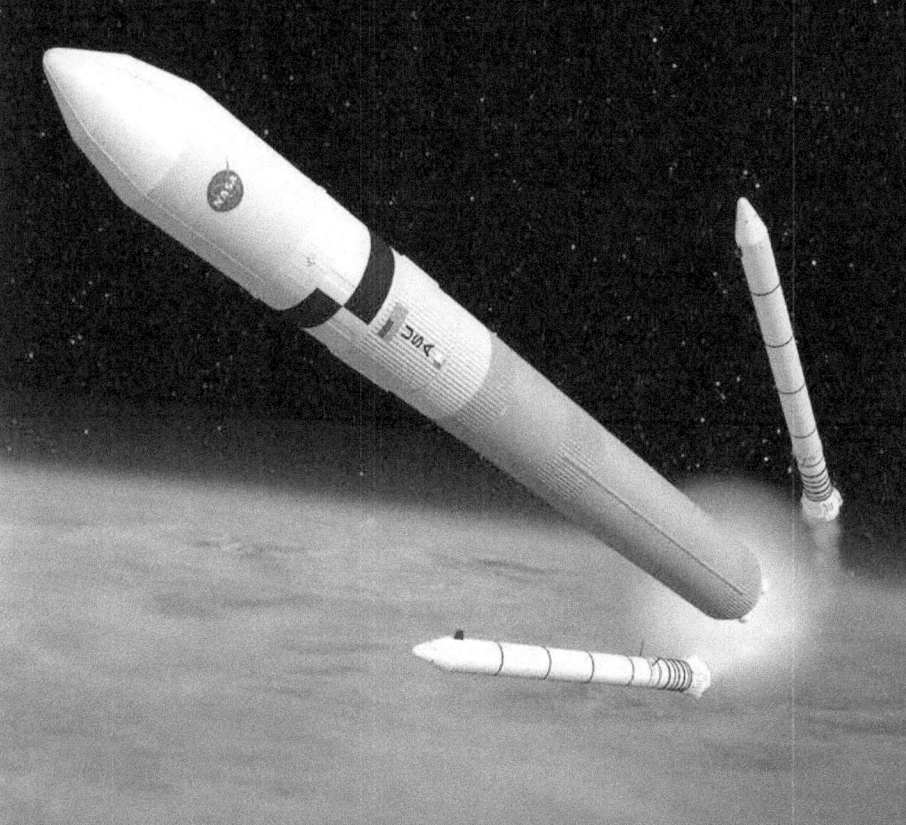

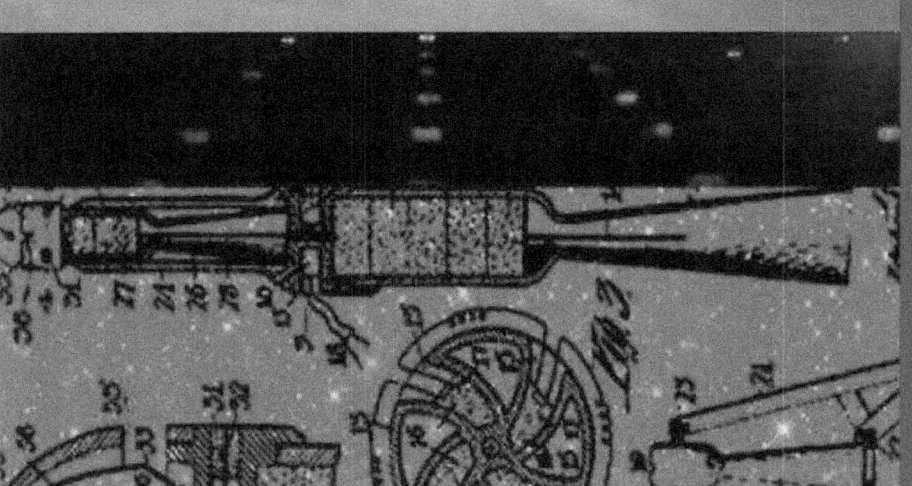

Educational Guide	
Educators	Grades K–12

EG-2007-12-179-MSFC

Acknowledgements

Adventures in Rocket Science is an expansion of the NASA guidebook *Rockets* by Deborah Shearer, Greg Vogt and Carla Rosenberg.

New and additional material was written and compiled by:

Vince Huegele, Marshall Space Flight Center Scientist and National Association of Rocketry (NAR) Education Chair

Kristy Hill, Education Specialist for WILL Technology, NASA Marshall Space Flight Center

Brenda Terry, Executive Director, Alabama Mathematics, Science and Technology Education Coalition (AMSTEC)

This document was produced through The NASA Explorer Institute (NEI)

Funding for the development of this rocketry guidebook for Informal Education was provided by NASA Explorer Institute and directed by the Marshall Space Flight Center Academic Affairs Office 2008.

Table of Contents

Adventures in Rocket Science Activity Matrix v

Standards vi

How to Use this Guide vii

Adventures in Rocket Science Introduction: Flight of a Model Rocket 1

Practical Rocketry 7

A Brief History of Rockets 17

Rocket History Glossary 24

Constellation Program America's Fleet of Next Generation Launch Vehicles 27

Activities 31

    1. Move It! 32

    2. Magic Marbles 35

    3. Shuttle Drag Parachute 38

    4. Paper Rockets 43

    5. Straw Rockets 49

    6. Rocket Racer 50

    7. 3-2-1 POP! 58

    8. Newton Car 62

    9. Pop Can Hero Engine 68

   10. Rocket Transportation 74

   11. Parachute Area Versus Drop Time 77

   12. Balloon Staging 83

   13. Water Bottle Rocket Assembly 85

   14. The Nose Cone Experts 90

15.	Racing Against Friction	95
16.	The Parachuting Egg	100
17.	Egg Drop Lander	103
18.	At the Drop of a Ball	105
19.	Free Fall Rocket Ball Drop	107
20.	Altitude Tracking	109
21.	The Scale of a Model Rocket	117
22.	Rocket Motion Video Studies	120
23.	Rocket Parachute Flight Duration Study	123
24.	Predicting and Measuring Rocket Parachute Drift Rate	126
25.	Project Enterprise	128

Glossary		143
Appendix A – Assembly Instructions		145
Appendix B – Rocket Safety Code		155
Appendix C – Rocket Principles		159
Appendix D – Rocketry Resource Materials		165

Adventures in Rocket Science Activity Matrix

Grade Level Groups					Activity	Altitude	Velocity	Acceleration	Page Number
1-3	4-6	7-9	10-12						
🚀				1	Move It! (balloon direction)		X		32
🚀				2	Magic Marbles			X	35
🚀				3	Shuttle Drag Parachute			X	38
🚀	🚀			4	Paper Rockets	X		X	43
🚀	🚀			5	Straw Rockets	X			49
🚀	🚀			6	Rocket Racer		X		50
🚀	🚀			7	3-2-1 POP!	X	X	X	58
🚀	🚀	🚀		8	Newton Car			X	62
🚀	🚀	🚀		9	Pop Can Hero Engine				68
🚀	🚀	🚀		10	Rocket Transportation	X			74
	🚀			11	Parachute Area Versus Drop Time			X	77
	🚀	🚀		12	Balloon Staging		X		83
	🚀	🚀	🚀	13	Water Bottle Rocket Assembly	X	X	X	85
		🚀		14	The Nose Cone Experts	X			90
		🚀		15	Racing Against Friction		X		95
		🚀		16	The Parachuting Egg			X	100
		🚀		17	Egg Drop Lander				103
		🚀	🚀	18	At the Drop of a Ball	X		X	105
		🚀	🚀	19	Free Fall Rocket Ball Drop	X			107
			🚀	20	Altitude Tracking	X			109
			🚀	21	The Scale of a Model Rocket		X	X	117
			🚀	22	Rocket Motion Video Studies		X		120
			🚀	23	Predicting and Measuring Rocket Parachute Drift Rate				123
			🚀	24	Rocket Parachute Flight Duration			X	126
	🚀	🚀	🚀	25	Project Enterprise (2 weeks)	X	X	X	128

Legend: X : Denotes the primary target word focus (altitude, velocity and/or acceleration).
▒ : Denotes secondary target word.
🚀 : Denotes suggested grade level.

Standards

Activity	Science as Inquiry	Physical Science	Position and Motion of Objects	Properties of Objects and Materials	Unifying Concepts and Processes	Change, Consistency and Measurement	Evidence, Models and Explanation	Science and Technology	Abilities of Technological Design	Understanding Science and Technology
Move It! (balloon direction)	X	X	X							
Magic Marbles	X	X	X	X	X		X			
Shuttle Drag Parachute	X	X	X			X				
Paper Rockets	X	X	X		X	X	X	X	X	
Straw Rockets	X	X	X		X	X	X	X	X	
Rocket Racer	X	X	X		X	X		X		
3-2-1 POP!		X	X					X	X	X
Newton Car	X	X		X	X	X	X			
Pop Can Hero Engine	X	X	X		X	X		X		X
Rocket Transportation	X	X	X					X	X	
Parachute Area Versus Drop Time	X	X	X	X	X	X	X		X	X
Balloon Staging		X	X					X	X	X
Water Bottle Rocket Assembly		X	X					X	X	
The Nose Cone Experts	X	X		X	X		X			
Racing Against Friction	X	X		X	X	X	X			
The Parachuting Egg	X	X	X	X	X	X	X		X	
Egg Drop Lander	X	X	X	X	X	X	X		X	
At the Drop of a Ball	X	X					X			
Free Fall Rocket Ball Drop	X	X					X			
Altitude Tracking		X	X					X	X	X
The Scale of a Model Rocket		X				X	X	X	X	X
Rocket Motion Video Studies	X	X	X			X	X	X	X	X
Predicting and Measuring Rocket Parachute Drift Rate	X	X	X	X		X	X			
Rocket Parachute Flight Duration	X	X	X	X	X		X	X		
Project Enterprise (2 weeks)	X	X	X					X	X	

How to Use This Guide

Acknowledgements

The excitement and thrill from experiencing the launch of rocket-powered vehicles ignites engagement and imagination in children as well as adults. Constructing and launching model rockets certainly provided a charge of excitement among a group of National Aeronautics and Space Administration (NASA) and higher education professionals as they planned an Exploration Systems Mission Directorate (ESMD) sponsored K-12 rocketry curriculum and mentor/coach workshop.

This guide was prepared as a tool useful for informal education venues (4-H, Boys and Girls Clubs, Boy Scouts, Girl Scouts, etc.), science clubs and related programs, and can be adopted for formal education settings. An exciting and productive study in rocket science can be implemented using the selected activities for the above-mentioned settings. The guide's activities can be correlated to meet formal education's core curriculum objectives or used for ancillary enrichment and extension activities.

The series of activities in this guide require basic, inexpensive materials. They demonstrate fundamental principles of rocket science. The activities were selected and organized around the target concepts of altitude, velocity and acceleration (i.e., height, distance and speed). The Adventures in Rocket Science Activity Matrix on page v organizes the activities by suggested grade level bands as well as target concepts.

Examine the matrix for a grade level band (e.g., grades 1–3). Select a target word located in that grade level band (e.g., velocity). The activity, "Move It!" will address the concept of velocity. Activities are included within each grade level band to address the three target concepts. Some activities can be used to address multiple target words. The length of time for each activity is approximately one or two hours with the exception of the Project Enterprise activity which can take up to two weeks to complete.

The guide includes a compilation of NASA-developed and originally developed activities that emphasize hands-on involvement, prediction, data collection and interpretation, teamwork, and problem solving. Project background and an edited, student-friendly version of *The History of Rocketry* is included. The guide includes an appendix with construction directions for various project activities, safety guidelines and a glossary of terms.

Adventures in Rocket Science Introduction: Flight of a Model Rocket

Guidebook Background

The NASA Explorers Institute (NEI) funded the development of this model rocketry guidebook for educators. Directed by the Marshall Space Flight Center (MSFC) Academic Affairs Office, education specialists created the curriculum with knowledgeable amateur rocketeers who are members of the National Association of Rocketry (NAR). The material was tested in a workshop pairing NAR members with informal educators from science centers, 4-H clubs, Girl Scout troops and after school programs. One purpose of the workshop was to introduce groups to the use of the guidebook with students in informal settings. The activities have been very successful in engaging and teaching students and are enjoyable for the teachers.

The Rocketry Education Pipeline

The theme of this book is that rocketry can be used to teach physical science beginning at any level. The subject has successive lessons, like a K through 12 pipeline course, that can lead a student to a technical career. Students begin an orientation to motion and velocity by launching air straw rockets and pump-up water bottle rockets. Then, they are introduced to force and acceleration as demonstrated in model rockets. In middle school, students can apply their understanding in design contests and move on in high school to launching payload experiments in larger models. Model rocketry is fully representative of the scientific process and the engineering applications of real world challenges. Rocketry studies from high school can be continued on a college level until the student emerges from the pipeline into the job market. This guidebook presents the flow of model rocket science activities that support the pipeline.

The Marvel of Motion

The study of motion is one of the fundamentals of physics that students should learn. The dynamic properties of an object are given by its altitude, velocity and acceleration. From these properties come the concepts of measurement and rate of change, which are the foundation of physical science. Rocketry demonstrates these concepts in a way that students on many grade levels can understand and remember.

One way to explain motion is to think about what is happening when you throw a ball. The ball is accelerated as the arm moves until the hand releases it; then, the ball appears to travel at a constant velocity from the pitcher to the catcher, though in actuality the thrown baseball is decelerating slightly due to air resistance, or drag, after the ball is released by the pitcher. When tossing a ball straight up, gravity immediately starts slowing the ball down, until it stops at the top of its flight path and then falls down at increasing speed, again due to the effect of gravity.

Instead of a ball, the rocket activities in this guide can be used to better illustrate this simple acceleration and free-fall motion. The first worksheet entitled The "Flight of a Water Bottle Rocket," page 3, shows the different points of change in the water bottle rocket flight. As the students launch the rockets, have them watch the flights and then answer the questions on the worksheet on page 4. You may prompt them before the flights to be observing for "When does the rocket go the fastest?" and "When is its speed changing the most?" After they choose answers, compare those to the worksheet answers and learn how altitude, velocity and acceleration are related to each other in flight.

The model rocket flight chart is different from the water rocket in that instead of the rocket free falling from apogee (its highest point), it pops a parachute or streamer to slow down its descent. The "Flight of a Model Rocket" worksheet builds on the water rocket profile to teach another kind of motion.

The rocket's altitude, velocity and acceleration and their relation in time can be illustrated in graphs. The plots of the relation of altitude, velocity and acceleration of a rocket on page 6 show how the values of the respective aspects change from ascent to apogee and on to landing. A model rocket simulation program will give a similar plot for the rocket being tested. The plots of altitude, velocity and acceleration are actually derived through calculus, so students progressing in math may get their first gentle exposure here.

Tell the students the flight profile of a model rocket with a parachute is the same as for the Solid Rocket Boosters (SRB) on the Space Shuttle and the Ares V. The SRBs launch, ascend, burn out, reach apogee, then deploy parachutes to land in order to be recovered and reloaded. The main differences between a NASA rocket and a model rocket is that a large rocket continuously controls its position during ascent, using moveable engine nozzles and aerodynamic surfaces (such as fins); whereas, a model rocket does not. The physics principles and even the flight phases are the same. If students understand the fundamentals of how a model rocket works, they will understand how any rocket works.

The first rockets ever built, the fire-arrows of the Chinese, were not very reliable. Many just exploded on launching. Others flew on erratic courses and landed in the wrong place. Being a rocketeer in the days of the fire-arrows must have been an exciting, but also a highly dangerous activity.

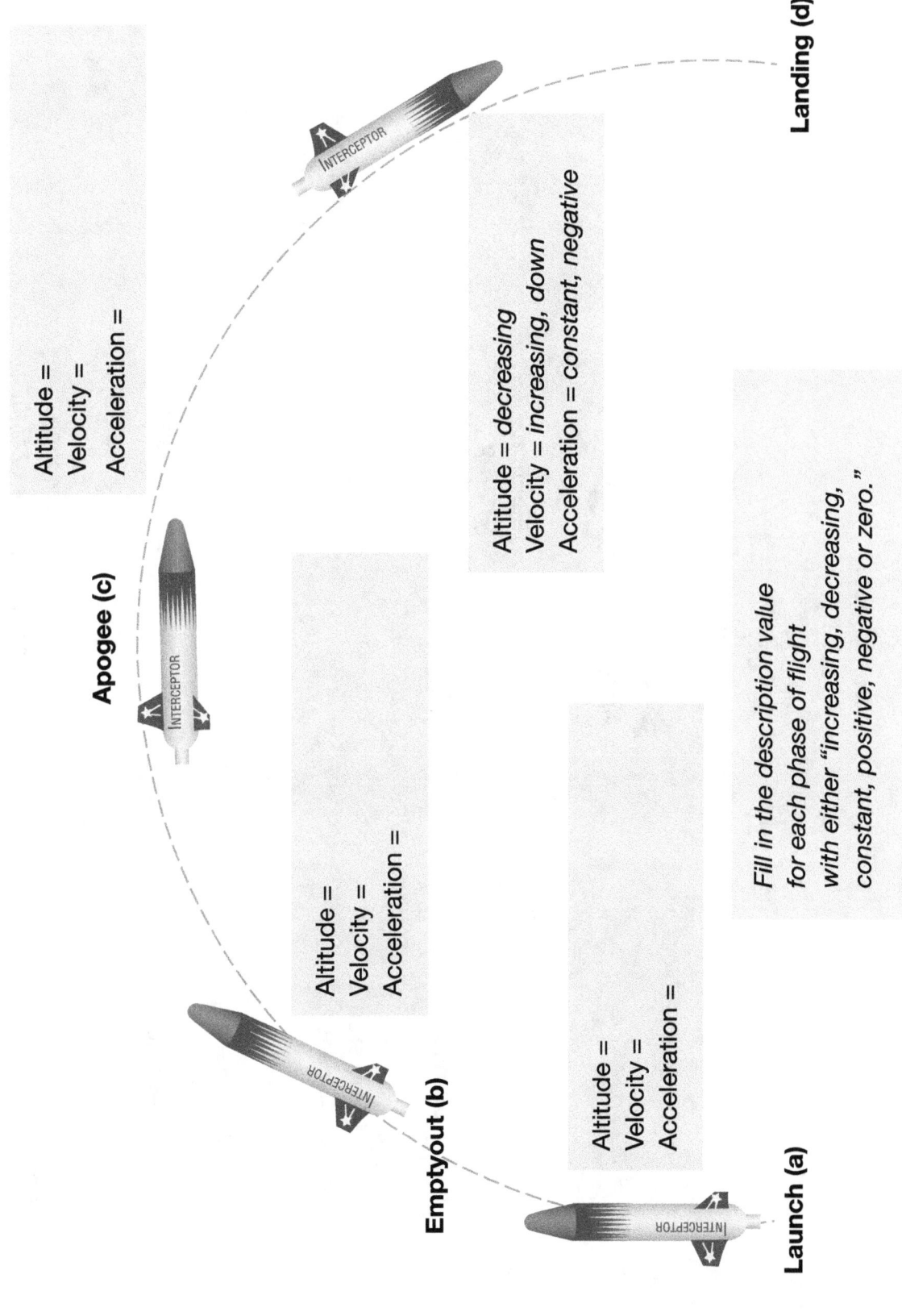

Flight of a Model Rocket

Worksheet

Altitude =
Velocity =
Acceleration =

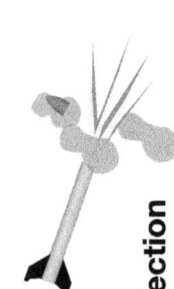

Apogee

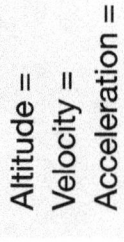

Ejection Charge

Parachute Descent

Altitude =
Velocity =
Acceleration =

Recovery

Fill in the description for each phase of flight

Altitude =
Velocity =
Acceleration =

Coasting Flight

Burnout

Powered Ascent

Altitude =
Velocity =
Acceleration =

Launch

Flight of the Water Bottle Rocket
Worksheet Answers

Apogee (c)
Altitude = *maximum*
Velocity = *zero*
Acceleration = *constant, negative*

Altitude = *increasing*
Velocity = *decreasing up*
Acceleration = *constant, negative*

Emptyout (b)

Altitude = *decreasing*
Velocity = *increasing, down*
Acceleration = *constant, negative*

Altitude = *increasing*
Velocity = *increasing up*
Acceleration = *constant, positive*

Launch (a)   **Landing (d)**

Flight of the Model Rocket
Worksheet Answers

Altitude = *maximum*
Velocity = *zero*
Acceleration = *constant, negative*

Apogee

Altitude = *increasing*
Velocity = *decreasing up*
Acceleration = *constant, negative*

Coasting Flight   **Ejection Charge**

Burnout

Powered Ascent

Parachute Descent

Altitude = *increasing*
Velocity = *increasing, up*
Acceleration = *constant, positive*

Altitude = *decreasing*
Velocity = *constant, down*
Acceleration = *zero*

Launch

Recovery

5

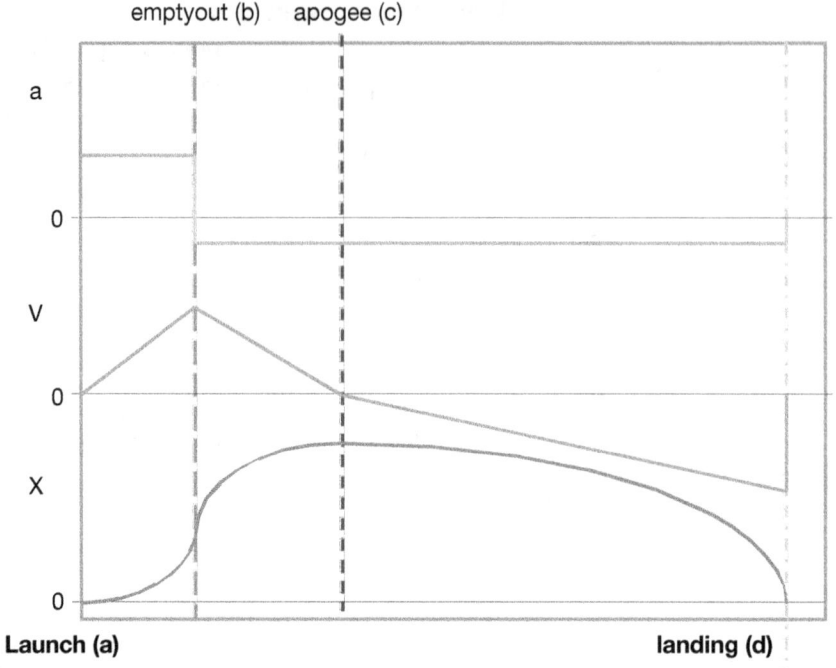
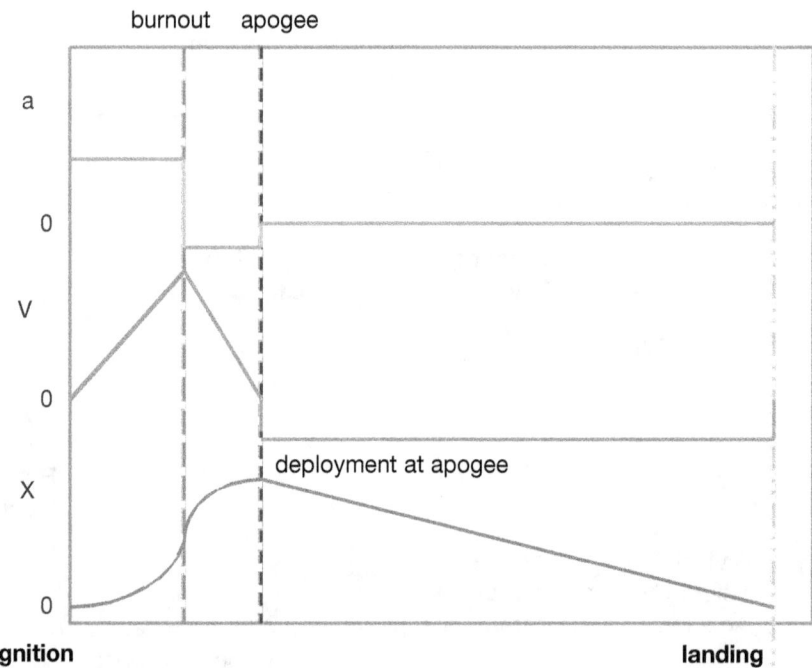

Practical Rocketry

Today, rockets are much more reliable. They fly on precise courses and are capable of going fast enough to escape the gravitational pull of the Earth. Modern rockets are also more efficient today because we have an understanding of the scientific principles behind rocketry. Our understanding has led us to develop a wide variety of advanced rocket hardware and devise new propellants that can be used for longer trips and more powerful take offs.

Rocket Engines and Their Propellants
Most rockets today operate with either solid or liquid propellants. The word propellant does not mean simply fuel, as you might think; it means both fuel and oxidizer. The fuel is the chemical the rocket burns but, for burning to take place, an oxidizer (typically oxygen) must be present. Jet engines draw oxygen into their engines from the surrounding air. Rockets do not have the luxury that jet planes have; they must carry the oxidizer with them into space where there is no air.

In the following discussion about rocket propulsion, several terms appear to be used synonymously, but actually have distinct meanings. These terms are rocket, rocket motor and rocket engine. In simple terms, a rocket is a vehicle that contains everything needed to place a payload into space. Rocket motors and rocket engines are propulsion devices that produce the thrust necessary to lift a rocket into space. A rocket motor is a simple device that converts stored rocket propellant into hot gases to produce thrust, such as a solid rocket motor. A rocket engine is a more complicated machine with moving parts that converts stored rocket propellants into hot gases to produce thrust. An example of a rocket engine is a liquid-propellant engine, which is a machine with valves and often pumps that performs the energy conversion to generate thrust.

Solid rocket propellants, which are dry to the touch, contain both the fuel and oxidizer combined together in the chemical mixture itself. Usually the fuel is a mixture of hydrogen compounds and carbon, and the oxidizer is made up of oxygen compounds.

Liquid propellants are kept in separate containers, one for the fuel and the other for the oxidizer. The fuel and oxidizer are mixed together in the engine to produce thrust.

A rocket with a liquid propulsion system can be very complex, but a rocket with a solid motor is much simpler because the solid motor propulsion system consists of a nozzle, a case, insulation, propellant and an igniter. The case of the engine is usually a relatively thin metal that is lined with insulation to keep the propellant from burning through. The propellant itself is packed inside the insulation layer, which is packed inside the motor casing.

Many solid-propellant rocket motors feature a hollow core that runs through the propellant. Motors that do not have the hollow core must be ignited at the lower end of the propellant, and burning proceeds from one end of the motor to the other. In all cases, only the surface of the propellant burns. However, to get higher thrust, the hollow core is used. This increases the surface area of the propellants available for burning. The propellants burn from the inside out at a much higher rate, sending mass out the nozzle at a higher rate and speed. This typically results in greater thrust. Some propellant cores are star shaped to increase the burning surface even more. Typical solid motor propellants are compressed black powder (used in Estes model rocket motors) and ammonium perchlorate composite propellant (used in the Space Shuttle SRBs and higher end model rocket motors).

To ignite solid propellants, many kinds of igniters can be used. Fuses were used to ignite fire-arrows, but sometimes these ignited too quickly and burned the rocketeer. A far safer and more reliable form of ignition used today in model rockets is one that employs electricity. An electric current, coming through wires from some distance away, heats up a special wire inside the rocket. The igniter raises the temperature of the propellant to the combustion point. Often, there is another pyrogenic mixture on the hotwire to more quickly reach the combustion temperature of the propellant.

Other igniters, especially those for large rockets, are often small rocket motors themselves. The small motor inside the hollow core blasts a stream of flames through the propellant core to ignite the inner surface area very quickly.

The nozzle in a solid-propellant motor is an opening at the back of the rocket that permits the hot expanding gases to escape. The narrow part of the nozzle is the throat. Just beyond the throat is the exit cone.

The purpose of the nozzle is to increase the acceleration of the gases as they leave the rocket and thereby maximize the thrust. It does this by cutting down the opening through which the gases can escape. To see how this works, you can experiment with a garden hose that has a spray nozzle attachment. This kind of nozzle does not have an exit cone, but that does not matter in the experiment. The important point about the nozzle is that the size of the opening can be varied to change the thrust.

Start with the opening at its widest point. Watch how far the water squirts and feel the thrust produced by the departing water. Now reduce the diameter of the opening, and again note the distance the water squirts and feel the thrust. Rocket nozzles work the same way.

As with the inside of the rocket motor case, insulation is often needed to protect the nozzle from the hot gases. The usual insulation is one that gradually erodes as the gas passes through. Small pieces of the insulation get very hot and break away from the nozzle. As they are blown away, heat is carried away with them. As will be discussed later, there are other methods of protecting the nozzle and combustion chamber materials from excess heat.

One of these methods is to use one of the propellants to cool the nozzle and combustion chamber by sending the

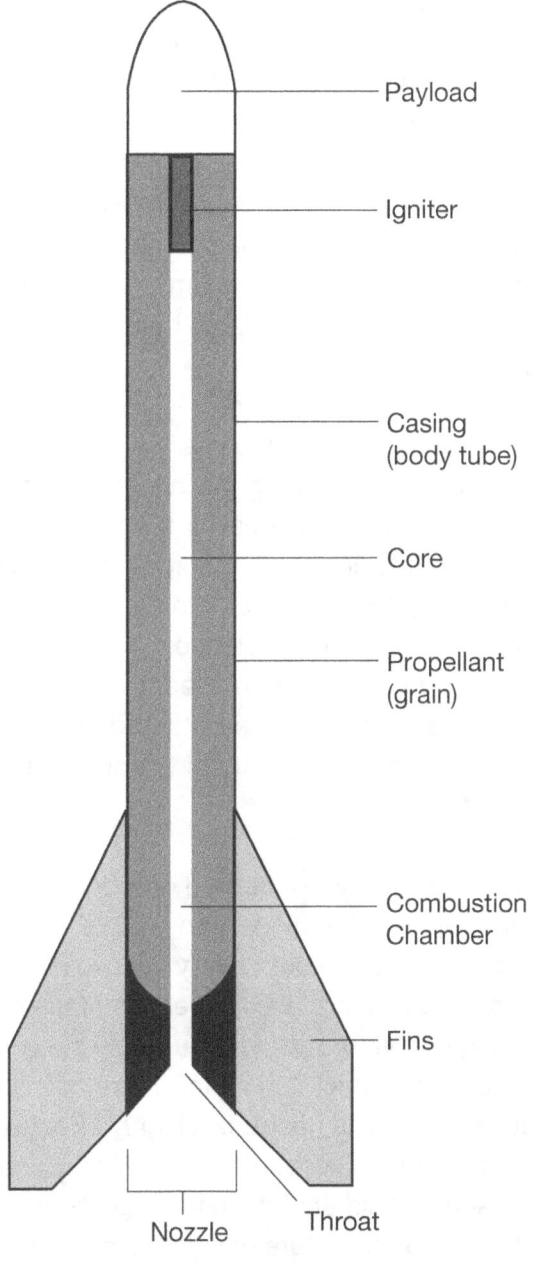

Solid Propellant Rocket

easier said than done when you consider that combustion temperatures may reach 6,000°F (3,315.56 °C), which exceeds the melting point of most metals that are suitable for use in rocket engines and motors. Fuel must often be sprayed directly onto the inner wall of the combustion chamber to help cool the chamber wall. Spraying fuel onto the chamber wall to keep the temperature of the chamber wall lower may seem counterintuitive, but it works because the fuel evaporates on the chamber wall, pulling heat out of the metal before the fuel itself is burned.

The other main kind of rocket engine is one that uses liquid propellants that may be either pumped or fed into the engine by pressure. This is a much more complicated engine, as is evidenced by the fact that solid rocket engines were used for at least 700 years before the first successful liquid engine was tested.

Contrasted with a relatively simple solid rocket motor, a liquid propellant engine requires a complicated propulsion system. This propulsion system must have propellant tanks (one for the fuel and one for the oxidizer), pressurant tanks to push the liquid propellant out of the tanks and into the engine, and valves to control the flow of the propellant to the engines. The liquid engine itself has valves and sometimes a pump, in addition to the combustion chamber and nozzle that you would see in a solid rocket motor.

propellant through small passages within the nozzle and chamber wall before sending the propellant into the combustion chamber. This is called regenerative cooling.

Another method to protect the chamber and nozzle from the heat of combustion is to use materials that can survive the heat for the life of the rocket. However, this is much

The fuel of a liquid-propellant rocket is usually kerosene or liquid hydrogen; the oxidizer is usually liquid oxygen. They are combined inside a cavity called the combustion chamber. Here, the propellants burn and build up high temperatures and pressures, and the expanding gas escapes through the nozzle at the lower end.

Because the combustion operates best under high pressures, the propellants need to be forced into the thrust chamber. Modern liquid rockets often use powerful, lightweight turbine pumps to take care of this job. Another method to increase the pressure within the combustion chamber is to use a high-pressure gas (typically, helium or nitrogen) to pressurize the propellants before they are injected into the engine.

With any rocket, and especially with liquid-propellant rockets, weight is an important factor. In general, the heavier the rocket, the more thrust is needed to get it off the ground. However, getting a rocket into space is a balancing act. You must balance the mass of the rocket and payload to be lifted into space, the efficiency of the propellant and engines, and the overall complexity of the propulsion system.

One especially good method of reducing the weight of rocket engines is to make the exit cone of the nozzle out of very lightweight metals. However, the extremely hot, fast-moving gases that pass through the cone would quickly melt thin metal. Therefore, a cooling system is needed. A highly effective, though complex, cooling system that is used with some liquid engines takes advantage of the low temperature of liquid hydrogen. Hydrogen becomes a liquid when it is chilled to −423.4 °F (−253 °C). Before injecting the hydrogen into the combustion chamber, it is first circulated through small tubes that lace the walls of the exit cone. In a cutaway view, the exit cone wall looks like the edge of corrugated cardboard. The hydrogen in the tubes absorbs the excess heat entering the cone walls and prevents it from melting the walls away. It also makes the hydrogen more energetic because of the heat it picks

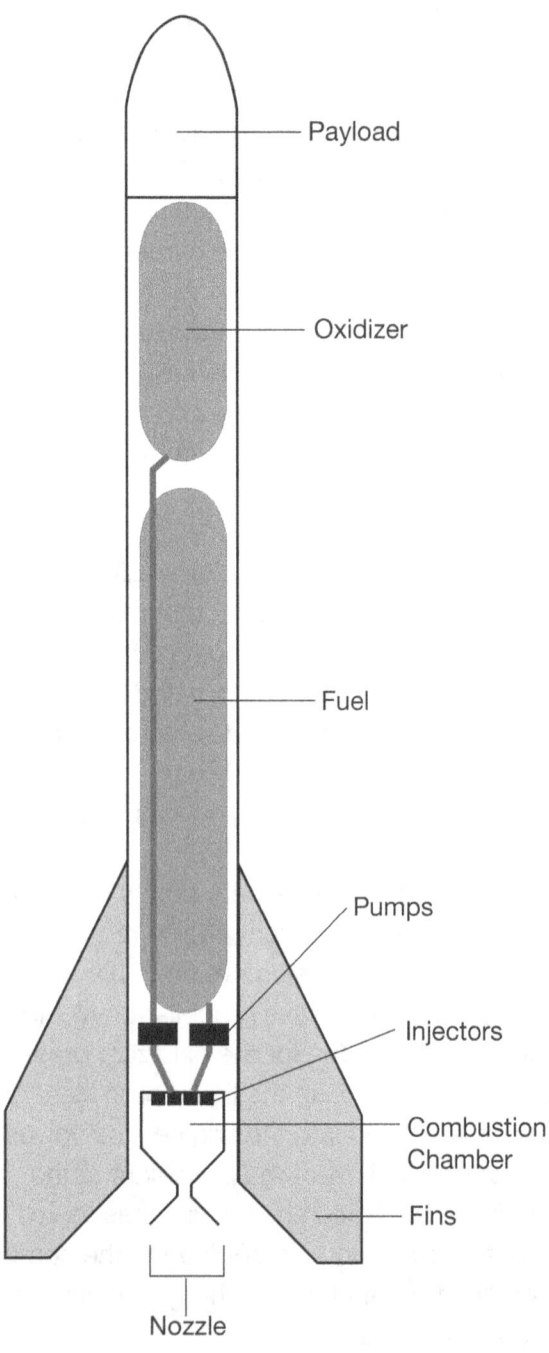

Liquid Propellant Rocket

To get the most power from the propellants, they must be mixed as completely as possible. Small injectors (nozzles) on the roof of the chamber spray and mix the propellants at the same time.

up from the combustion chamber wall. We call this kind of cooling system regenerative cooling. Typically, a regeneratively cooled rocket engine uses the fuel propellant to take the excess heat out of the combustion chamber wall before the fuel itself is consumed in combustion.

Engine Thrust Control
Controlling the thrust of an engine is very important for launching payloads (cargoes) into space. Thrusting for too short or too long of a period of time will cause a satellite to be placed in the wrong orbit. This could cause it to go too far into space to be useful or make the satellite fall back to Earth too soon. Thrusting in the wrong direction or at the wrong time will result in a similar situation.

A computer in the rocket's guidance system determines when thrust is needed and turns the engine on or off appropriately. Liquid engines do this by simply starting or stopping the flow of propellants into the combustion chamber. On more complicated flights, such as going to the Moon, the engines must be started and stopped several times, and multiple sets of engines are required to escape Earth's gravity.

Some liquid-propellant engines control the amount of engine thrust by varying the amount of propellant that enters the combustion chamber. Typically, the engine thrust varies for controlling the acceleration experienced by astronauts or to limit the aerodynamic forces on a vehicle.

Solid-propellant rockets are not as easy to control as liquid rockets. Typically, once started, the propellants burn until they are gone. They are very difficult to stop or slow down after the solid motor has ignited.

Sometimes fire extinguishers are built into the engine to stop the rocket in flight. However, using them is a tricky procedure and does not always work. Some solid-fuel engines have hatches on their sides that can be cut loose by remote control to release the chamber pressure and terminate thrust.

The burn rate of solid propellants is carefully planned in advance. The hollow core running the length of the propellants can be made into a star shape. At first, there is a very large surface available for burning, but as the points of the star burn away, the surface area is reduced. For a time, less of the propellant burns, and this reduces thrust. The Space Shuttle uses this technique to reduce vibrations early in its flight into orbit.

Note: Although most rockets used by governments and research organizations are very reliable, there is still great danger associated with the building and firing of rocket engines. Individuals interested in rocketry should never attempt to build their own engines unless supervised by a propulsion expert. Even the simplest-looking rocket engines are very complex. Case-wall bursting strength, propellant packing density, nozzle design and propellant chemistry are all design problems beyond the scope of most amateurs. Many home-built rocket engines have exploded in the faces of their builders with tragic consequences.

Stability and Control Systems
Building an efficient rocket engine is only part of the problem in producing a successful rocket. The rocket must also be stable in flight. A stable rocket is one that flies in a smooth, uniform direction. An unstable rocket flies along an erratic path, sometimes tumbling or changing direction. Unstable

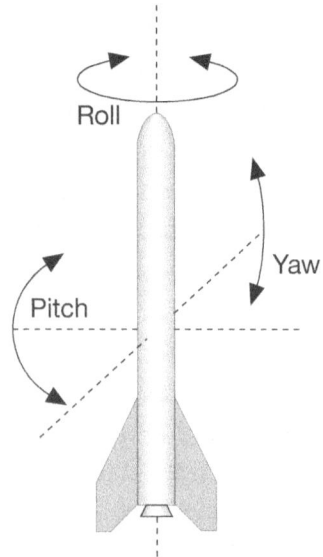

rockets are dangerous because it is not possible to predict where they will go. They may even turn upside down and suddenly head back directly to the launch pad or at the spectators.

Making a rocket stable requires some form of control system. Controls can be either active or passive. The difference between these and how they work will be explained later. It is first important to understand what makes a rocket stable or unstable.

All matter, regardless of size, mass or shape, has a point inside called the center of mass (CM). The CM is the exact spot where all of the mass of that object is perfectly balanced. You can easily find the CM of an object such as a ruler by balancing the object on your finger. If the material used to make the ruler is of uniform thickness and density, the CM should be at the halfway point between one end of the stick and the other. If the ruler was made of wood and a heavy nail was driven into one of its ends, the CM would no longer be in the middle. The balance point would then be nearer the end with the nail.

The CM is important in rocket flight because it is around this point that an unstable rocket tumbles. As a matter of fact, any object in flight tends to tumble. Throw a stick, and it tumbles end over end. Throw a ball, and it spins in flight. The act of spinning or tumbling is a way of becoming stabilized in flight. A Frisbee will go where you want it to only if you throw it with a deliberate spin. Try throwing a Frisbee without spinning it. If you succeed, you will see that the Frisbee flies in an erratic path and falls far short of its mark. Likewise, a knuckleball, which has little or no spin, follows an erratic flight from the pitcher to the catcher, making it very difficult for the batter to hit the ball.

In flight, spinning or tumbling takes place around one or more of three axes. They are called roll, pitch and yaw. The point where all three of these axes intersect is the CM. For rocket flight, the pitch and yaw axes are the most important because any movement in either of these two directions can cause the rocket to go off course. The roll axis is the least important because movement along this axis will not affect the flight path. In fact, a rolling motion will help stabilize the rocket in the same way a properly passed football is stabilized by rolling (spiraling) it in flight. Although a poorly passed football may still fly to its mark even if it tumbles rather than rolls, a rocket will not. The action-reaction energy of a football pass will be completely expended by the thrower the moment the ball leaves the hand. With rockets, thrust from the engine is still being produced while the rocket is in flight. Unstable motions about the pitch and yaw axes will cause the rocket to leave the planned course. To prevent this, a control system is needed to prevent or at least minimize unstable motions.

In addition to CM, there is another important center inside the rocket that affects its flight. This is the center of pressure (CP). The CP exists only when air is flowing past the moving rocket. This flowing air, rubbing and pushing against the outer surface of the rocket, can cause it to begin moving around one of its three axes. Think for a moment of a weather vane. A weather vane is an arrow-like stick that is mounted on a rooftop and used for telling wind direction. The arrow is attached to a vertical rod that acts as a pivot point. The arrow is balanced so that the CM is right at the pivot point. When the wind blows, the arrow turns, and the head of the arrow points into the oncoming wind. The tail of the arrow points in the downwind direction.

The reason that the weather vane arrow points into the wind is that the tail of the arrow has a much larger surface area than the arrowhead. The flowing air imparts a greater force to the tail than the head, and therefore the tail is pushed away. There is a point on the arrow where the surface area is the same on one side as the other. This spot is called the CP. The CP is not in the same place as the CM. If it was, the wind would favor neither end of the arrow, and the arrow would not point. The CP must be between the CM and the tail end of the arrow for the weather vane to point into the wind. This means that the tail end has more surface area than the head end.

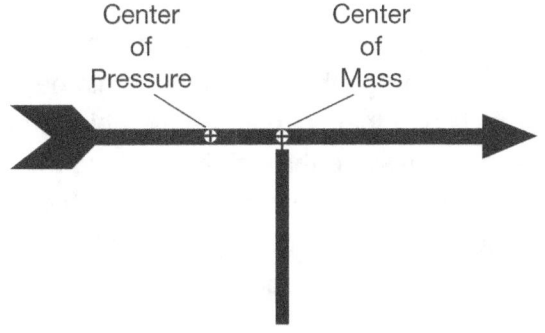

It is extremely important that the CP in a rocket be located toward the tail and the CM mass be located toward the nose. If they are in the same place or very near each other, then the rocket will be unstable in flight. The rocket would then try to rotate about the CM in the pitch and yaw axes, producing an erratic flight path. With the CP behind the center of gravity, the rocket will maintain a stable trajectory, or flight path.

Control systems for rockets are intended to keep a rocket stable in flight and to steer it. Small rockets usually require only a stabilizing control system. A large rocket, such as the ones that launch satellites into orbit, requires a system that not only stabilizes the rocket, but also enables it to change course while in flight.

Controls on rockets can either be passive or active. Passive controls are fixed devices that keep rockets stabilized by their very presence on the rocket's exterior. Active controls can be moved while the rocket is in flight to stabilize and steer the craft.

The simplest of all passive controls is a stick. The Chinese fire-arrows were simple rockets mounted on the ends of sticks. The stick kept the CP behind the CM. In spite of this, fire-arrows were notoriously inaccurate. Before the CP could take effect, air had to be flowing past the rocket. While still on the ground and immobile, the arrow might lurch and fire the wrong way.

Years later, the accuracy of fire-arrows was improved considerably by mounting them in a trough aimed in the proper direction. The trough guided the arrow in the right direction until it was moving fast enough to be stable on its own.

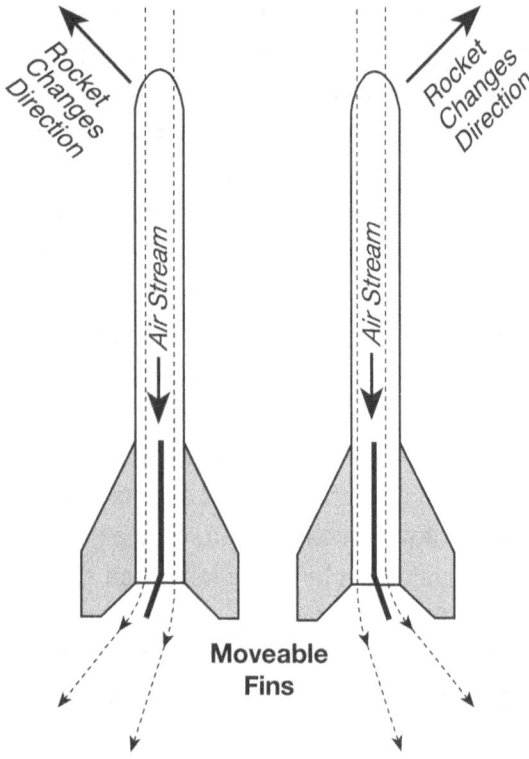

Moveable Fins

As will be explained in the next section, the weight of the rocket is a critical factor in performance and range. The fire-arrow stick added too much dead weight to the rocket and, therefore, limited its range considerably.

An important improvement in rocketry came with the replacement of sticks by clusters of lightweight fins mounted around the lower end near the nozzle. Fins could be made out of lightweight materials and be streamlined in shape. They gave rockets a dart-like appearance. The large surface area of the fins easily kept the CP behind the CM. Some experimenters even bent the lower tips of the fins in a pinwheel fashion to promote rapid spinning in flight. With these "spin fins," rockets become much more stable in flight. However, the increased drag created by the spin fins reduced the rocket's range.

With the start of modern rocketry in the 20th century, new ways were sought to improve rocket stability and at the same time reduce overall rocket weight. The answer to this was the development of active controls. Active control systems included vanes, movable fins, canards, gimbaled nozzles, vernier rockets, fuel injection and attitude-control rockets. Tilting fins and canards are quite similar to each other in appearance. The only real difference between them is their location on the rocket. Canards are mounted on the front end of the rocket, while the tilting fins are at the rear. In flight, the fins and canards tilt like rudders to deflect the air flow and cause the rocket to change course. Motion sensors on the rocket detect directional changes, and corrections can be made by slightly tilting the fins and canards. The advantage of these two devices is size and weight. They are smaller and lighter and produce less drag than large fins. Moveable fins also provide better pointing accuracy of the rocket than do fixed, immoveable fins.

Other active control systems can eliminate fins and canards altogether. By tilting the angle at which the exhaust gas leaves the rocket engine, course changes can be made in flight. Several techniques can be used for changing exhaust direction.

Vanes are small finlike devices that are placed inside the exhaust of the rocket engine. Tilting the vanes deflects the exhaust, and, by action-reaction, the rocket responds by pointing the opposite way.

Another method for changing the exhaust direction is to gimbal the nozzle. A gimbaled nozzle is one that is able to sway while exhaust gases are passing through it. By tilting the engine nozzle in the proper direction, the rocket responds by rotating about the center of gravity to change course.

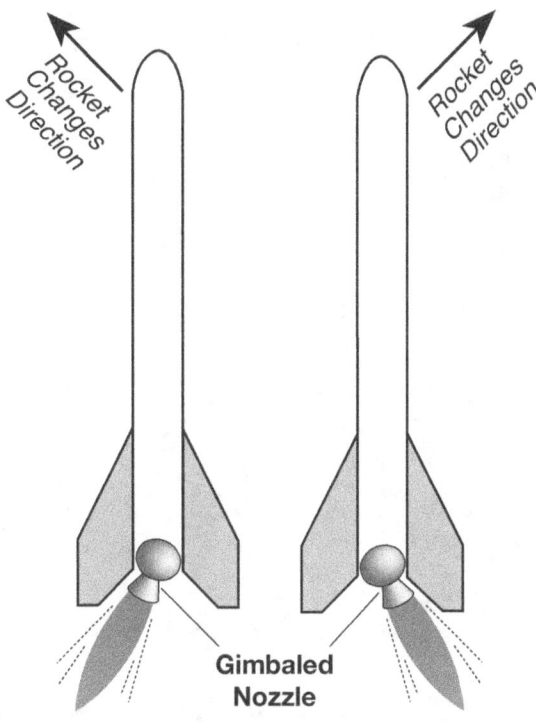

Gimbaled Nozzle

Vernier rockets can also be used to change direction. These are small rockets mounted on the outside of the large engine. When fired, they also produce a desired course change.

In space, only by spinning the rocket along the roll axis or by using active controls involving the engine exhaust can the rocket be stabilized or have its direction changed. Without air, fins and canards have nothing to work upon. (Science fiction movies showing rockets in space with wings and fins are long on fiction and short on science.) While coasting in space, the most common kinds of active controls used are attitude-control rockets. Small clusters of engines are mounted all around the vehicle. By firing the right combination of these small rockets, the vehicle can be turned in any direction. As soon as the spacecraft is aimed properly, the main engines fire, sending the rocket off in the new direction.

Mass

Mass is another important factor affecting the performance of a rocket. The mass of a rocket can make the difference between a successful flight and just wallowing around on the launch pad. As a basic principle of rocket flight, it can be said that for a rocket to leave the ground, the engine must produce a thrust that is greater than the total mass of the vehicle. It is obvious that a rocket with a lot of unnecessary mass will not be as efficient as one that is trimmed to just the bare essentials.

For an ideal rocket, the total mass of the vehicle should be distributed following this general formula: of the total mass, 91 percent should be propellants; 3 percent should be tanks, engines, fins, etc.; and 6 percent can be the payload.

Payloads may be satellites, astronauts, or spacecraft that will travel to other planets or moons. In determining the effectiveness of a rocket design, rocketeers speak in terms of mass fraction (MF). The mass of the propellants of the rocket divided by the total mass of the rocket gives MF.

The MF of the ideal rocket given above is 0.91. From the MF formula, one might think that an MF of 1.0 is perfect, but then the entire rocket would be nothing more than a lump of propellants that would simply ignite into a fireball. The larger the MF number, the smaller payload the rocket can carry; the smaller the MF number, the less its range becomes. An MF number of 0.91 is a good balance between payload-carrying capability and range. The Space Shuttle has an MF of approximately 0.82. The MF varies between the different orbiters in the Space Shuttle fleet and with the different payload weights of each mission.

Large rockets, able to carry a spacecraft into space, have serious weight problems. To reach space and proper orbital velocities, a great deal of propellant is needed; therefore, the tanks, engines and associated hardware become larger. Up to a point, bigger rockets can carry more payload than smaller rockets, but when they become too large, their structures weigh them down too much, and the MF is reduced to an impossible number.

A solution to the problem of giant rockets weighing too much can be credited to the 16th-century fireworks maker Johann Schmidlap. Schmidlap attached small rockets to the top of big ones. When the large rocket was exhausted, the rocket casing was dropped behind, and the remaining rocket fired. Much higher altitudes were achieved by this method. The rockets used by Schmidlap were called step rockets. (The Space Shuttle follows the step rocket principle by dropping off its SRBs and external tank when they are exhausted of propellants.) Today this technique of building a rocket is called staging. Thanks to staging, it has become possible not only to reach outer space, but the Moon and other planets too.

A Brief History of Rockets

Isaac Newton came up with three laws that govern our world. These laws are the Laws of Motion. They deal with position, velocity and acceleration. Position is an explanation of where something is, based on a certain origin (or starting place). For instance, you could say your teacher is in front of you in the classroom, but it is also true that he/she is in front of the chalkboard. We used you as the origin in the first position and the chalkboard in the second. As you can see, a description of location is always based on the origin. Now, velocity is the speed and direction something is moving. If I said your teacher is walking at 3 mph (4.83 kmph), I have given you his/her speed, but not velocity because I have not told you the direction in which he/she is traveling. Now, if I said that your teacher is moving towards the chalkboard at 3 mph (4.83 kmph) that would be a velocity because I have given you her speed, 3 mph (4.83 kmph), and her direction, towards the board. Acceleration is how velocity changes with time. For instance, if I said he/she was accelerating across the room at 3 mph^2 (4.83 $kmph^2$), then that would mean that for every hour he/she was traveling, his/her velocity would increase by another 3 mph (4.83 kmph). So, if his/her initial velocity was 3 mph, her acceleration was 3 mph^2 (4.83 $kmph^2$), and she traveled for 2 hours, what would her new velocity be?

Newton's first law states that an object traveling at a certain velocity (velocity can be zero) will remain traveling at that velocity as long as there are no forces acting on it. That makes sense, since our shoes do not start moving across the floor by themselves. They only move if we push them or pick them up.

Newton's second law states that an object will experience acceleration if a force is applied to this object and that the acceleration is in the direction of the applied force. Furthermore, the equation Force = mass $\times$ acceleration (F = m $\times$ a) governs this acceleration from an applied force. If you push your shoes with a constant force, they will accelerate according to the formula. Note that mass is not the same as weight. Mass is how much material a body has, usually measured in kilograms. Weight is the force a mass experiences in a certain gravity like that of Earth or on the Moon. Your weight on the Moon is less, but your mass is the same.

Newton's third law is that every action has an equal and opposite reaction. When we push against a wall, the wall pushes the exact same amount against our hands. Otherwise, one of us would begin moving according to the second law. Since it is equal and opposite, the net force is zero, so there is no acceleration. This law explains why it hurts when we fall down. Our body hits the ground with a force equal to our weight, which on Earth is approximately 10 times our mass. When we hit the ground, the ground hits us back, and that is why it hurts so badly.

Newton's laws can be applied to our everyday lives and experiences. They govern anything we do and cannot do for that matter. Rockets make use of Newton's Laws in order to carry out the tasks they are designed to do. Can you think of other examples that demonstrate the three Laws of Motion?

Rockets have been around for centuries. These rockets started off in a very simple form, like our fireworks, to become the rockets that we use for warfare and traveling into outer space. Much of rocketry science has been discovered within the last century, including new fuels, ways of controlling rockets, and new uses for rockets.

We will start our discussion of rocketry far back in time with the Greeks and continue until we reach modern day rocketry.

A Greek by the name of Hero invented a rocket-like device called an aeolipile. It used steam as a propulsive gas. Hero mounted a sphere on top of a water kettle. A fire below the kettle turned the water into steam, and the gas traveled through pipes to the sphere. Two L-shaped tubes on opposite sides of the sphere allowed the steam to escape and, in doing so, gave a thrust to the sphere that caused it to rotate. The aeolipile used the action-reaction principle that was not to be stated as a scientific law, by Newton, until the 17th century.

Would you believe the Chinese made rockets out of bamboo?

Although we know that inventions that used ideas related to rocketry were around for many centuries, we are uncertain as to when true rockets came into existence. One idea is that the Chinese invented the first rockets. The Chinese made a simple form of gunpowder from saltpeter, sulfur and charcoal dust that they put inside bamboo tubes and tossed into a fire for religious or celebration purposes. Some of the bamboo tubes may not have fully exploded, but instead shot out of the fire. Whatever the case, we do know that the Chinese began to shoot arrows with rockets attached in order to go longer distances, and they soon learned that the rockets would launch the arrows by themselves without bows.

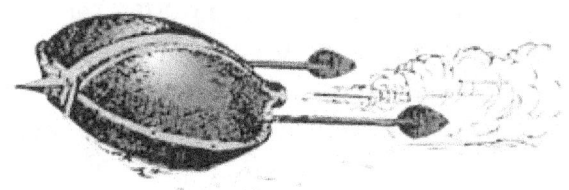 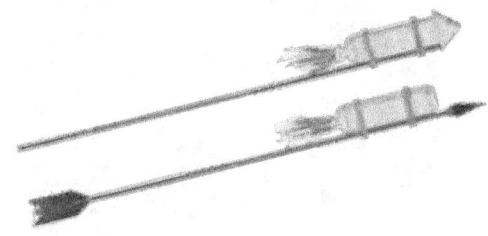

These rockets were first used in warfare against the Mongols in 1232 during the Battle of Kai-Keng. The Mongols referred to these weapons as "arrows of flying fire." The fire-arrows were a simple form of solid-propellant rocket. They used a bamboo tube filled with gunpowder and capped at one end. The other end was left open, and the tube was attached to an arrow. When the gunpowder was ignited, fire, smoke, and gas all escaped out the open end, which propelled the arrow towards the enemy. The stick of the arrow provided a guidance system to shoot it a general direction as it flew through the air. There is no way of knowing how many arrows hit their target, but the idea of "fire arrows" flying towards the Mongols must have disturbed them greatly. Soon after the Battle of Kai-Keng, the Mongols produced their own rockets and may have spread this valuable new invention to the European continent.

Between the 13th and 15th centuries, many rocket experiments were conducted in Europe. In England, Roger Bacon, a monk, improved gunpowder in order to increase the range of rockets. In France, Jean Froissart discovered that accuracy greatly increased by using tubes to launch rockets. Modern day bazookas are a result of this discovery. Joanes de Fontana of Italy designed a type of torpedo, or surface-running rocket, that sped across the top of the water and would set enemy ships on fire. By the time the 16th century rolled around, rockets fell out of use as weapons, but still remained in use as fireworks. Around this time, Johann Schmidlap, a German, invented the step rocket, which enabled fireworks to fly to higher altitudes. Basically, this multistage rocket was a large rocket that carried a smaller rocket inside. When the large rocket burned out, the smaller one continued flying before exploding and sending the beautiful shower of sparks we are accustomed to seeing. This idea of multistage rockets is the basic design for all rockets that go into outer space today.

The 17th century brought about scientific inquiry that laid the foundations for modern rocketry. The scientist Sir Isaac Newton made these discoveries between 1642 and 1727. Newton discovered the Three Laws of Motion. These laws explain why rockets work and how they are able to work in a vacuum, like outer space.

Newton's laws gave scientists new ideas on rocket design. Scientists experimenting with rockets began to shoot rockets that were more than 100 lb (45.36 kg). Some of these rockets were even so powerful that they bore deep holes into the ground at lift-off.

At the end of the 18th century and beginning of the 19th century, rockets soon began to be used as weapons of war once again. The people of India repelled the British in 1792 and in 1799 by using rockets. Colonel William Congreve, a British artillery expert in the battle, set out to design rockets for the British military as a result of fighting against India.

Congreve's rockets were very successful in battle. You may have even heard of their reputation. The rockets were used by British ships against Fort McHenry in the War of 1812, where they inspired Francis Scott Key to write "the rockets' red glare" words in a poem, which later became "The Star-Spangled Banner." Still, the rockets had not improved in accuracy. Their devastating nature in warfare was the fact that thousands of rockets could be fired at the enemy, not the power or accuracy of those rockets. Rockets were continuously used in warfare with great outcomes all over Europe; however, Austria came up against a Prussian weapon that surpassed the effectiveness of other rockets.

The Prussians had breech-loading cannons with rifled barrels and exploding warheads. After everyone realized that rockets were not the most effective weapons, their primary use once again became peacetime occurrences.

Around 1898, Konstantin Tsiolkovsky (1857–1935), a Russian schoolteacher, suggested the idea of space travel by rocket. He published a report in 1903 where he suggested liquid propellant instead of solid propellant to achieve greater height. He stated that the speed and height of a rocket were only limited by the exhaust velocity of the escaping gases. Because of his ideas, research and vision, Tsiolkovsky is known as the Father of Modern Astronautics.

In the early 1900s, Robert H. Goddard (1882–1945), an American, conducted experiments with rockets. He was trying to reach heights even greater than lighter-than-air balloons. He produced a pamphlet in 1919 entitled "A Method of Reaching Extreme Altitudes." It was a mathematical analysis, which today is called the meteorological sounding rocket.

Goddard made several important conclusions. Two of these were that rockets operate better in a vacuum than in air, and that multi-stage or step rockets would be the best way to achieve the altitude and velocity needed to escape Earth's gravity. Many did not believe Goddard on the basis that they thought propellants needed air to push against, which is not true.

Goddard first worked with solid-propellant rockets. In 1915, he experimented with different types of solid fuels and measured the exhaust velocities of the burning gases. While working with these rockets, he became convinced that a liquid propellant would be much better; however, no one had successfully used a liquid-propellant rocket before.

The task of building a liquid-propellant rocket was very difficult due to the need for fuel and oxygen tanks, turbines, and combustion chambers. Still, Goddard built the first successful liquid-propellant rocket on March 16, 1926. The rocket was fueled by oxygen and gasoline, flew for only 2.5 s, and went only 41.01 ft (12.5 m) up, and out 183.73 ft (56 m) away into a cabbage patch.

Although it does not compare to today's standards, Goddard's gasoline rocket was the beginning of a whole new era in rocketry. Goddard continued his experiments for many years. His rockets got bigger and flew higher. He developed a gyroscope system for flight control, a compartment to put scientific instruments in, and a parachute recovery system to return rocket and instruments safely to Earth. For these achievements, Goddard is known as the Father of Modern Rocketry.

Herman Oberth (1894–1989) of Germany published a book about rocket travel into outer space in 1923. Because of this book, many rocket societies sprang up all around the world. In Germany, a rocket society by the name of Verein fur Raumschiffahrt (Society for Space Travel) created the V-2 rocket that was used against London in World War II. In 1937, German rocket scientists and engineers assembled on the shores of the Baltic Sea.

They built and flew the most advanced rocket of its time under the leadership of Wernher von Braun.

The V-2 rocket was small by comparison to today's rockets. It burned a mixture of liquid oxygen and alcohol at a rate of about 1 ton (907.18 kg) every 7 s. Once launched, the V-2 could destroy whole city blocks. The V-2 was too late to change the outcome of the war. We found out after the war had ended that the Germans had already laid plans for advanced missiles that could span the Atlantic Ocean and land in the United States (U.S.). However, these rockets would have had small payload capacities, not causing as much destruction as those for closer range fighting. After the war, many German rocket scientists moved to the U.S. and the Soviet Union.

The U.S. and the Soviet Union realized the potential of rocketry as a military weapon and began large-scale experimental programs. The U.S. started with high-altitude atmospheric sounding rockets, one of Goddard's early ideas. Afterwards, many different kinds of medium- and long-range intercontinental ballistic missiles were created. This was the beginning of the U.S. space program. Missiles from this program, such as the Redstone, Atlas and Titan, would launch astronauts into space.

Would you believe that Russian astronauts are called Cosmonauts?

On October 4, 1957, the Soviet Union launched the first Earth-orbiting artificial satellite called Sputnik I. This satellite was the first successful entry in the race for space between the superpower nations. Less than a month later, the Soviets launched a satellite with a dog, named Laika, on board. Laika lived for seven days in outer space before being put to sleep before oxygen ran out.

On January 31, 1958, the U.S. finally launched a satellite of its own, Explorer I. In October of 1958, the U.S. organized the space program by officially creating the National Aeronautics and Space Administration (NASA). NASA's goal was the peaceful exploration of space for the benefit of all humankind.

As of today, many people and machines have been launched into space. Astronauts have orbited Earth and landed on the Moon and robot spacecrafts have traveled to other planets to explore their surfaces.

Space is now open to exploration and commercialization. Satellites have enabled scientists to run experiments to investigate our world, forecast weather and communicate quickly and efficiently around the world. As our needs grow, so do the payloads that are taken out to space. As a result, a wide array of powerful and versatile rockets have been and will be built.

As you can see, our universe has been opened up for exploration by our growing understanding of rocketry. If it were not for those who experimented with the gunpowder rockets centuries ago and those who can take modern day rockets and apply new ideas, we would not be able to explore the universe and all its glory.

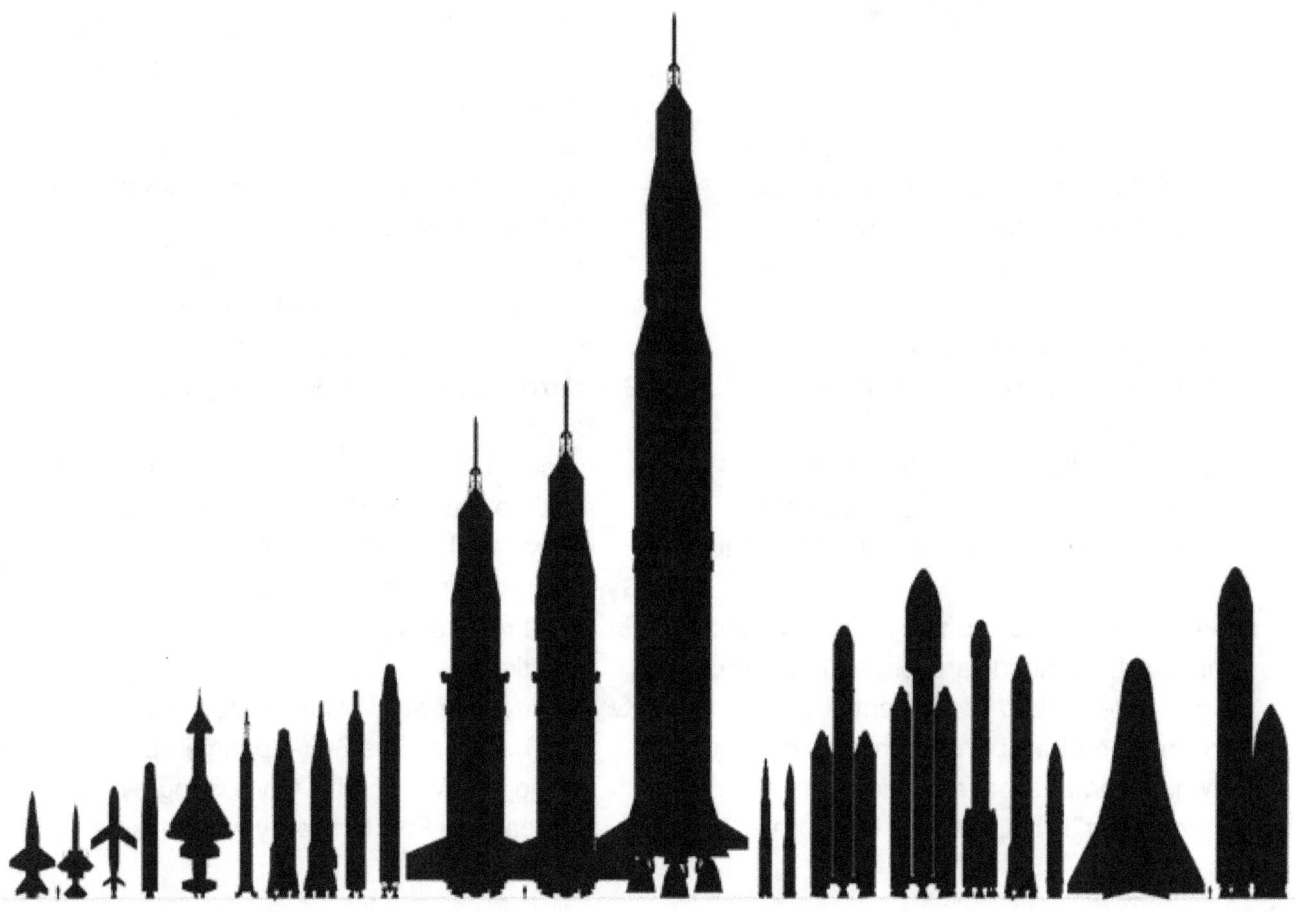

Rocket History Glossary

Aeolipile: An apparatus consisting of a closed globe with one or more projecting bent tubes through which steam is made to pass from the globe, causing it to revolve.

Bacon, Roger: An English monk who improved gunpowder, which increased the range of rockets.

Battle of Kai-Keng: A battle between the Chinese and Mongolians in 1232 A.D. where rockets were first used for military purposes.

Breech-loading cannons: A type of cannon in which warheads are loaded into the backside instead of the front.

Combustion chamber: An enclosure in which combustion, especially of a fuel or propellant, is initiated and controlled.

Congreve, Colonel William (1772–1828): Artillery expert who developed rockets after a battle with India; led the attack that inspired Francis Scott Key.

Exhaust velocity: The average actual velocity at which exhaust material leaves the open end of a rocket engine.

Explorer I: First satellite the U.S. successfully sent into orbit around the Earth.

Fontana, Joanes de: Inventor from Italy who designed the surface running rocket or torpedo.

Fort McHenry: Located in Baltimore, Maryland, it is best known for its role in the War of 1812, when it successfully defended Baltimore Harbor from the British navy. It was during this bombardment of the fort that Francis Scott Key was inspired to write the Star-Spangled Banner.

Froissart, Jean: Inventor who discovered that by using tubes you could greatly increase the accuracy of fired rockets.

Goddard, Robert H. (1882–1945): A pioneer in the field of rocketry, during his time he was ridiculed for his ideas, which were ahead of his time. After his death, he would become known as the "Father of Modern Rocketry."

Gyroscope: A mounted wheel that spins so that it resists movement or change of direction.

Key, Francis Scott: Wrote the poem that later became "The Star-Spangled Banner" during the War of 1812 while Congreve was bombarding Fort McHenry with rockets.

Laika: One of the Russian space dogs and the first living passenger in orbit around Earth.

Liquid-propellant rocket: A rocket that is fueled by liquid propellant such as liquid oxygen and gasoline.

Mongols: An ethnic group that originated from modern day Mongolia and parts of Russia and China. For more information, read about Genghis Khan.

NASA: National Aeronautics and Space Administration created by the U.S. Government in 1958, responsible for the U.S. space program and aerospace research. It is a civilian organization that does both civilian and military aerospace research.

Newton, Sir Isaac (1642–1727): English mathematician and physicist remembered for developing calculus, his law of gravitation and his three laws of motion.

Oberth, Herman (1894–1989): Scientist who wrote about the possibility of space travel via rockets that inspired many rocket societies to spring up all over the world.

Rifled barrel: A barrel that has internal spiral grooves that cause the ammunition to spin as it leaves, making its path more controlled.

Rocket: A vehicle or device propelled by one or more engines or motors.

Rocket engine: A machine with moving parts that converts stored propellant into hot gases to produce thrust, such as liquid engines with valves and pumps.

Rocket motor: A simple device that converts stored propellant into hot gases to produce thrust, such as a solid rocket motor.

Saltpeter (KNO_3): A transparent white crystalline compound used to pickle meat and in the manufacture of pyrotechnics, explosives, matches, rocket propellants and fertilizers.

Schmidlap, Johann: German fireworks maker who was the first to use step rockets to achieve greater heights.

Solid-propellant rocket: A rocket that is fueled by solids, such as gunpowder or ammonium perchlorate.

Sputnik I: First Earth-orbiting artificial satellite launched by the Soviet Union on October 4, 1957.

Solid Rocket Booster (SRB): A motor composed of a segmented motor case loaded with solid propellants, an ignition system, a movable nozzle and the necessary instrumentation and integration hardware.

Staged rocket: A rocket that fires several different engines or sets of engines in stages, where the first stage lifts the rocket off the pad, then that stage falls away as the second stage engines ignite to carry the remaining rocket to an even higher altitude. More stages may be used to achieve an objective. For example, the Saturn V Moon Rocket used five stages to place astronauts on the Moon and an additional two stages to return back to Earth.

Step rocket: A rocket that uses multiple rocket engines, also called a step engine. When one rocket extinguishes, the other ignites sending the rocket even further. This is the basic set up of modern day rockets that go into outer space.

Sulfur: A yellow nonmetallic element that is used in gunpowder.

Tsiolkovsky, Konstantin (1857–1935): A Russian school teacher who was the first to conceive of space travel by rockets, suggested the use of liquid propellant instead of solid. He is known as the Father of Modern Astronautics.

Turbine: Rotary engine in which the kinetic energy of a moving fluid is converted into mechanical energy by causing a bladed rotor to rotate.

V-2 rocket: An early ballistic missile used by Germany during the later stages of World War II against mostly British and Belgian targets.

Vacuum: An absence of matter, a void, or a state of emptiness.

Verein fur Raumschiffahrt: An association of amateur rocket enthusiasts in Germany (1927–1933) that consisted of engineers who would contribute to early spaceflight.

von Braun, Wernher: A German scientist and one of the leading figures of rocket technology in Germany and the U.S. He led the V-1 and V-2 rocket effort in Germany, worked in the U.S. after WWII through the secret effort called Operation Paperclip and later worked for NASA, and is known as the "Father of the U.S. Space Program."

Warhead: The front part of a guided missile or rocket or torpedo that carries the nuclear or explosive charge.

War of 1812: A war (1812–1814) between the U.S. and England where England was trying to interfere with American trade with France.

Body tube: The main body of the rocket that contains parachute and engine and to which the nose cone and fins are affixed.

Engine mount: This is how the force from the engine is transmitted to the rocket body to make it move.

Fins: These are on the back end or bottom of the rocket and help provide stability during flight.

Launch lug: These are small "straw-like" attachments to the rocket body through which a rod is placed in order to increase stability during launch.

Nose cone payload: The nose cone is at the top or front end of the rocket. It can either be empty or carry something such as weights or instruments (on larger models).

Parachute: This is a piece of material that helps the rocket slowly descend back to the Earth in order for a safe recovery to take place.

Parachute lines: How the parachute is connected to the rocket, usually connected to the nose cone.

Recovery wadding: A piece of cloth placed between the engine mount and the parachute in order to protect the parachute from being burned up when the ejection charge is ignited.

Shock cord: Connects the nose cone to the rocket body in order to keep the rocket together during recovery.

Solid rocket engine: The method of propellant for most model rockets. The engine is used once and is disposable. It can be purchased from hobby stores, is very stable at room temperature, and only ignites when exposed to high heat. It produces an exhaust gas that will shoot out of one end (the one facing the back of the rocket), propelling it forward

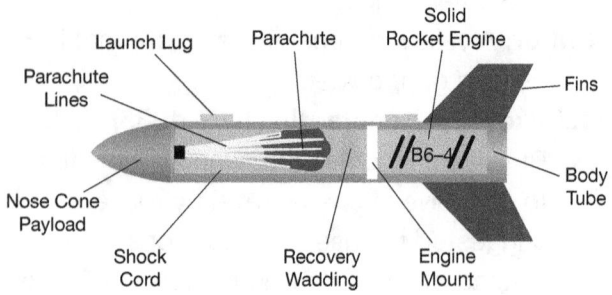

Constellation Program
America's Fleet of Next-Generation Launch Vehicles

The Ares I Crew Launch Vehicle
Ares I
NASA is developing hardware and systems for the Ares I rocket that will send future astronauts into orbit. Built on cutting-edge launch technologies, evolved powerful Apollo and Space Shuttle propulsion elements, and decades of NASA space flight experience, Ares I is the essential core of a safe, reliable, and cost-effective space transportation system; one that will carry crewed missions back to the Moon, on to Mars, and out to other destinations in the solar system.

Ares I has an in-line, two-stage rocket configuration topped by the Orion crew vehicle and its launch abort system. In addition to the vehicle's primary mission, carrying crews of four astronauts in Orion to rendezvous with the Ares V Earth Departure Stage (EDS) and Altair lunar lander for missions to the Moon, Ares I may also use its 26-ton (23,586.8-kg) payload capacity to deliver six astronauts or supplies to the International Space Station or to park payloads in orbit for retrieval by other spacecraft bound for the Moon or other destinations.

During launch, the first-stage booster powers the vehicle toward low-Earth orbit. In mid-flight, the booster separates and the upper stage's J-2X engine ignites, putting the vehicle into a circular orbit.

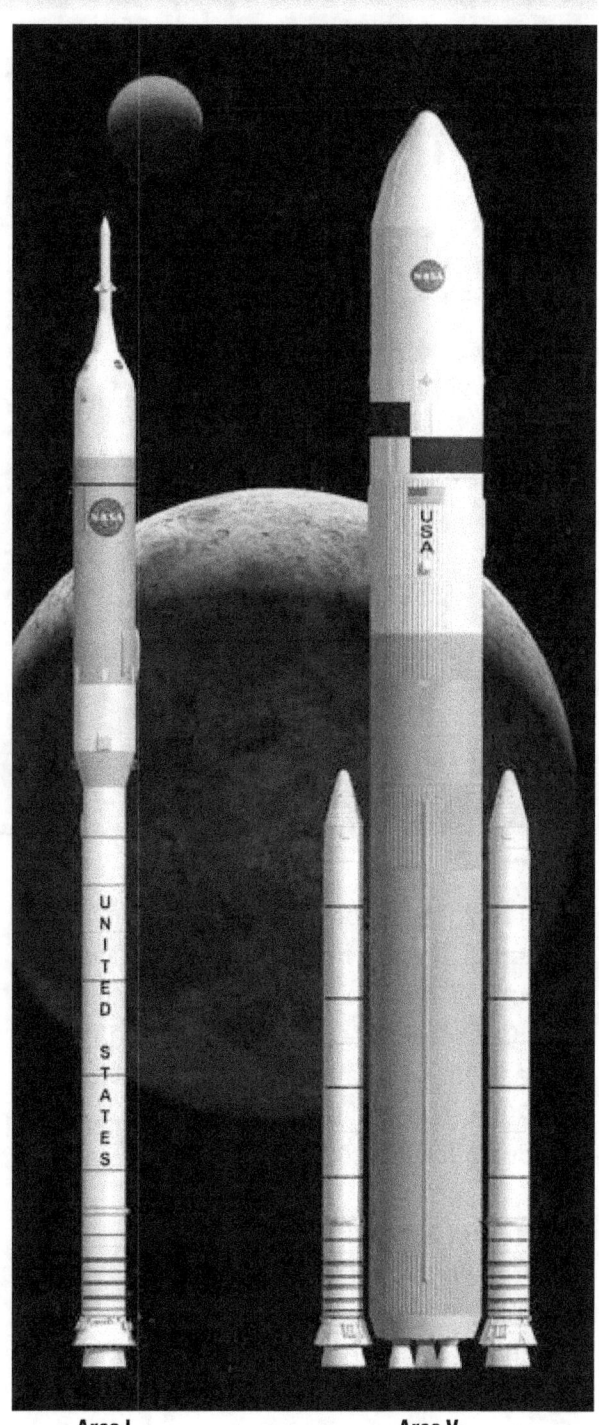

Ares I   Ares V

Apollo Saturn V | Space Shuttle | Ares I | Ares V

The first flight test of Ares I, called Ares I-X, will launch in 2009. The first crewed mission of Orion will be launched aboard Ares I in 2013. Crew transportation to the International Space Station is planned to begin no later than 2015. The first lunar excursion is scheduled for the 2020 timeframe.

Ares I First Stage
The Ares I first stage is a single, five-segment reusable solid rocket booster (SRB) derived from the Space Shuttle program's reusable solid rocket motor that burns a specially formulated and shaped solid propellant.

A newly designed forward adapter called a frustum and an interstage will mate the vehicle's first stage to the upper stage, and will be equipped with booster separation motors to disconnect the stages during ascent.

Ares I Upper Stage/Upper Stage Engine
The Ares I second or upper stage is propelled by a J-2X main engine fueled with liquid oxygen and liquid hydrogen.

The J-2X is an evolved variation of two historic predecessors: the powerful J-2 engine that propelled the Apollo-era Saturn IB and Saturn

V rockets, and the J-2S, a simplified version of the J-2 developed and tested in the early 1970s, but never flown.

The Ares V Cargo Launch Vehicle
Ares V

NASA is planning and designing hardware and propulsion systems for the Ares V cargo launch vehicle, the heavy lifter of America's next-generation space fleet.

During launch, the Ares V first stage and core propulsion stage power it upward toward Earth orbit. After separation from the spent core stage, the upper stage EDS takes over, and its J-2X engine puts the vehicle into a circular orbit.

The cargo vehicle's propulsion system can lift heavy structures and hardware to orbit or fire its engines for translunar injection, a trajectory designed to intersect with the Moon. Such lift capabilities will enable NASA to carry a variety of robust science and exploration payloads to space and could possibly take future crews to Mars and beyond.

The first launch of Ares V is scheduled for 2018. The first crewed mission to the Moon is scheduled to occur by 2020.

Ares V First Stage

The first stage of the Ares V vehicle comprises two, five-segment SRBs for liftoff. Derived from the space shuttle SRBs, they are similar to the single booster that serves as the first stage for the cargo vehicle's sister craft, Ares I.

Ares V Core Stage/Core Stage Engine

The first stage twin SRBs flank a single, liquid-fueled central booster element. With a diameter as wide as the Saturn V first stage, the central booster tank delivers liquid oxygen/liquid hydrogen fuel to five RS-68 rocket engines. The RS-68s used on Ares V will be a modified version of the

ones currently used in the U.S. Air Force's Delta IV evolved expendable launch vehicle program and commercial launch applications. The RS-68 engines serve as the core stage propulsion for Ares V.

An interstage cylinder is atop the central booster element. It includes booster separation motors and a newly designed forward adapter that mates the first stage with the EDS.

Ares V Earth Departure Stage/Engine
Like the Ares I upper stage, the Ares V EDS is propelled by a J-2X main engine fueled with liquid oxygen and liquid hydrogen.

The EDS separates from the core stage and its J-2X engine ignites midflight.

Once in orbit, the Orion, delivered to orbit by Ares I, docks with the orbiting EDS carrying the Altair lander, which will ferry astronauts to and from the Moon's surface. Once mated with the Orion crew module, the departure stage fires its engine to achieve escape velocity, the speed necessary to break free of Earth's gravity, and the new lunar vessel begins its journey to the Moon.

The EDS is then jettisoned, leaving Orion and the lunar lander mated. Once the four astronauts arrive in lunar orbit, they transfer to the lunar lander and descend to the Moon's surface. Orion remains in lunar orbit until the astronauts depart from the Moon in the lunar vessel, rendezvous with the crew module in orbit and return to Earth.

Lunar Lander
Anchored atop the EDS is a composite shroud protecting the Altair lunar lander.

This module includes the descent stage that will carry explorers to the Moon's surface and the ascent stage that will return them to lunar orbit to rendezvous with Orion, their ride home to Earth.

Activities

Move It!

Note: Excellent activity to be used prior to Balloon Staging and/or Rocket Transportation.

Objective
Learn how an airplane, rocket or shuttle moves forward.

- Target concept: Velocity
- Preparation time: 10 min
- Duration of activity: 45 min
- Student group size: Pairs

Materials and Tools
- Balloons
- Student Sheets

Background Information
Aircraft engines provide a constant source of thrust to provide the vehicle forward movement. Thrust is the force that moves an aircraft through the air. It is used to overcome the drag of an airplane and to overcome the weight of a rocket. Thrust is generated by the engines of the aircraft through some kind of propulsion system.

After introducing the scientific method, the teacher can walk the students through this simple experiment to help them become accustomed to using the Scientific Method.

The balloon moves in the opposite direction of the flow of the released air because every action has an opposite and equal reaction (Newton's Third Law). Since the air is released from one small hole, the release of the air is focused in one direction.

Procedure
1. Ask students:
   a. Why is an airplane able to move forward?
   b. Possible answers might be:
      i. Air coming out the back end pushes it forward.
      ii. The engines make it go.
2. Tell students that they are going to get to experiment to see how a jet or plane moves forward.
3. Distribute the student sheets and balloons.
4. Explain to the students that some of the steps have been done for them. They will complete the steps that have not been done. Read over each completed step.
5. As students follow the procedures, remind them to use all of their senses.

Discussion/Wrap-up
- After completion and cleanup, discuss the activity.

Move It! Data Sheet

Name: _____

Balloon Thrust Experiment Log Answer Sheet

1. State the problem. Question: What do I want to know?	
2. Form a hypothesis.	Air comes out the back end and pushes it forward. Air blows out one way, and the aircraft moves the other way.
3. Design an experiment. Materials and Procedures • What steps will I take to do this experiment? • What things do I need?	Materials • Balloon • Student Sheet • Pencil Procedure: 1. Collect materials. 2. Fill the balloon with air. do not tie a knot—hold it so that no air gets out. 3. Hold the balloon up with the opening facing left. 4. Let go of balloon. Write what happens. 5. Repeat steps 2–4, with the opening of the balloon facing to the right. 6. Repeat steps 2–4, with the opening of the balloon facing up. 7. Repeat steps 2–4, with the opeining of the balloon facing the ground.

Move It! Data Sheet

Name: _____

Balloon Thrust Experiment Log Answer Sheet

4. Perform the experiment. Observe and record the data What did I learn during this experiment?	Left: Right: Up: Down:
5. Organize and analyze data. In general, describe what happens when you release the balloon.	
6. Draw conclusions. What did I learn? Was my hypothesis right or wrong? Can I tell why?	

Magic Marbles

Objective
Observe Newton's First Law of Motion.
- Target concept: Acceleration
- Preparation time: 10 min
- Duration of activity: 45 min
- Student group size: Teams of two or three

Materials
(per group)
- Two rulers
- Several marbles
- Tape

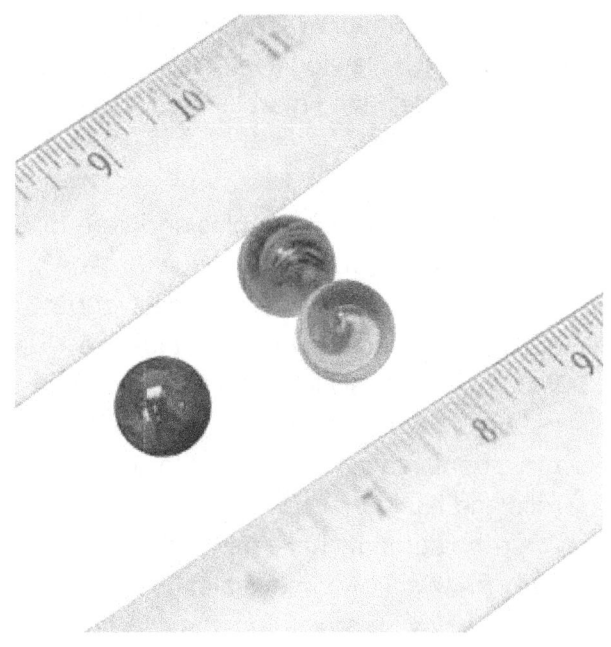

Management
You may set up test stations for younger students before beginning this lesson.

Background Information
It may be the most famous physics experiment ever done. Four hundred years ago, Galileo Galilei started dropping things off the Leaning Tower of Pisa and timing their falls. In those days, it was widely thought that heavy objects fell faster than light ones, but he found that everything hit the ground at the same time. For cannon balls, musket balls, gold, silver and wood, gravity accelerated each item downward at the same rate, regardless of mass or composition.

Today, this is called "universality of free fall" or the "equivalence principle," and it is a cornerstone of modern physics. Isaac Newton made his own tests of the principle using pendulums. Satisfied, he devised his famous laws of motion. Later, Einstein crafted his Theory of General Relativity, assuming the principle was true.

Newton's First Law states that every object will remain at rest or in uniform motion in a straight line unless compelled to change its state by the action of an external force. This is normally taken as the definition of inertia. The key point here is that if there is no net force acting on an object (if all the external forces cancel each other out), then the object will maintain a constant velocity. If that velocity is zero, then the object remains at rest. If an external force is applied, the velocity will change because of the force.

There are many excellent examples of Newton's First Law involving aerodynamics. The first law describes the motion of an airplane when the pilot changes the throttle setting of the engine. The motion of a ball falling down through the atmosphere and a model rocket being launched up into the atmosphere are both examples of Newton's First Law. The first law also describes the motion of a kite when the wind changes.

In this lesson, students will use marbles and rulers to observe Newton's First Law (See page 160 of this guide) using marbles and rulers.

Guidelines
1. Distribute the rulers, marbles and tape to each group. Show students how to tape the rulers to a flat surface, parallel, about a half-inch apart. Have them place the two marbles a few inches from each other between the two rulers. Have students take turns gently tapping one marble so that it rolls and hits the second one.
2. After the students have had a chance to perform this activity several times, have them put two marbles together, and then a third several inches away. Have them take turns gently tapping the single marble so that it hits the other two.
3. Allow students to experiment with other combinations.

Discussion
Ask the students the following questions:
- What happened in the first trial? (The marble that had been rolling stopped. The one that it hit started rolling. Explain to them that the force of the rolling marble moves to the stationary one. The stationary marble stops the first one, and then starts rolling.)
- What happened as you added more marbles? (The rolling marble stopped, the middle one stayed still and the third one rolled. Tell them that the movement of the first marble went through the second marble into the third.)
- What did you find out when using your own combinations? (The number of marbles pushed is how many were moved when hit.)

Reading Selection:
1. Read the corresponding NASA explores K-4 article, "Was Galileo Wrong" orally.
2. Discuss the current NASA project, Magic Marbles, and what students discovered.
3. Discuss Galileo's work and how it influenced other famous scientists. Reread the definitions of the laws of motion.
4. Tell the students, "Newton took Galileo's work and performed his own tests. From our experiment we saw what Newton discovered."
5. Relate to Newton's First Law (refer to Newton's laws in appendix C).

Was Galileo Wrong?

Galileo was a scientist who lived 400 years ago. In his time, people thought that heavy things fell faster than light ones. Galileo wanted to test this. So, he dropped different kinds of balls off a tall tower and found that mass did not matter. It did not matter what things were made of either. They all fell at the same speed.

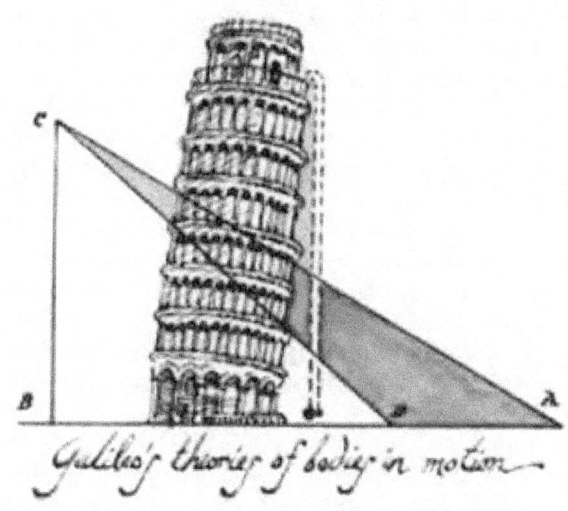

Lots of scientists have used what Galileo found. Isaac Newton used it to write his laws of motion, and Einstein used it to write science laws. But, what if Galileo's science was wrong? It would not be the first time it happened, and it will not be the last. Science is always changing. We keep learning new things. We find that old laws are not always right. People used to think that things in space moved around the Earth. Now we know that the Sun is the center of our universe. The more we know, the more we learn. That is why we explore, and space is one of the best places to do that.

A group at NASA is testing Galileo's ideas. They are using laser beams. The beams are shot toward the Moon. The crews that went to the Moon left mirrors there years ago. These were made to catch the beams. Then, they bounce them back to Earth. Lasers are used because their beams stay together as they go through space.

It is a new kind of tower experiment. Balls are not dropped, however. Instead, the team will watch as the Earth and Moon drop toward the Sun. The Earth and Moon are made of different things. They are not the same size. Are they moving toward the Sun at the same speed? We might have the answer soon. But will it be the answer we want? This is one case where being wrong may be a good thing.

Courtesy of NASA's ESMD
Published by NASA explores: September 23, 2004

Shuttle Drag Parachute

Objective
- Test the effects of a drag parachute on a Shuttle model and to calculate the speed variances.
- Target concept: Acceleration
- Preparation time: 10–30 min
- Duration of activity: 50–60 min
- Student group size: Teams of three to five students

Materials and Tools
(Per group)
- Small shoebox or a clean half-gallon-size milk carton
- Round balloon
- One piece of black construction paper
- Two pieces of white construction paper
- String or yarn
- One small plastic shopping bag with handles
- Tape
- Scissors
- Stopwatch or watch with a second hand
- Yardstick, rulers or meter stick

Management
This lesson is designed as a group project. For younger grades, the teacher will need to build the model and demonstrate the activity.

Background Information
Drag is the resistance to motion through the air. When the Space Shuttle comes back to Earth, it has no power once it comes into the atmosphere. This 178,000-lb (80,739.44-kg) craft lands by gliding to the ground. The Shuttle is equipped with a special feature called a speed brake to help slow it down. It is really not a brake, but it increases drag to slow the craft. This drag would be like driving down a road in a car and opening both doors. The increased resistance drops the speed quickly.

When the Shuttle decreases its speed to about 185 knots (343 km) per hour, it deploys a drag parachute to help slow it at an even faster rate. This type of drag parachute is also used on some high-performance jets.

In this lesson, students will build a Shuttle model and test how a drag parachute affects its speed.

Guidelines
1. Explain that drag is very important in slowing down the Shuttle. The best way to increase drag is to add space that air has to flow over. A parachute will slow down an object because the air fills it up, and it pulls back in the opposite direction from the air that is filling it. Then, the parachute pulls back on the object to which it is attached.

2. Model the following directions for each group while they build their Shuttle model. For older students, hand out the student sheets, and allow them to follow the directions while you observe their progress.
   a. Take a small shoebox or a half-gallon-size milk carton, and punch a hole in one end with a pencil.
   b. Place a balloon inside the shoebox or carton, and pull the open end through the hole. See the diagram below. This will be the back of the Shuttle.
   c. Make a nose cone for the Shuttle using the black construction paper. See the pattern below.
   d. Form the cone as shown in the following figure.

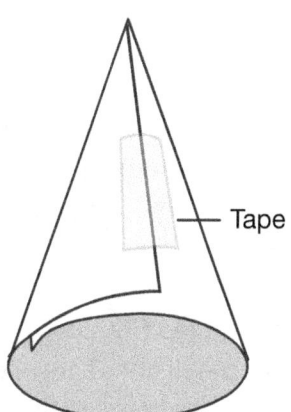

   e. Attach the cone to the opposite end of the box or carton from the balloon. This will be the front of the Shuttle.
   f. Make wings to attach to the sides of the Shuttle by drawing two large triangles on the white construction paper, and tape one to each side of the box with the larger end of the triangle positioned at the back of the Shuttle. See the following figure.

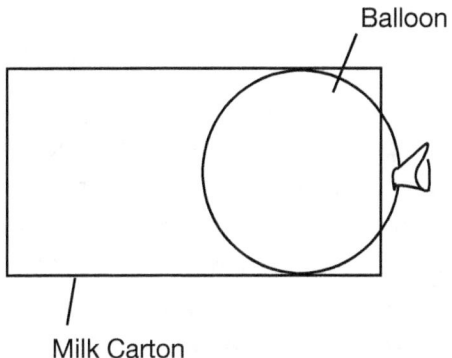

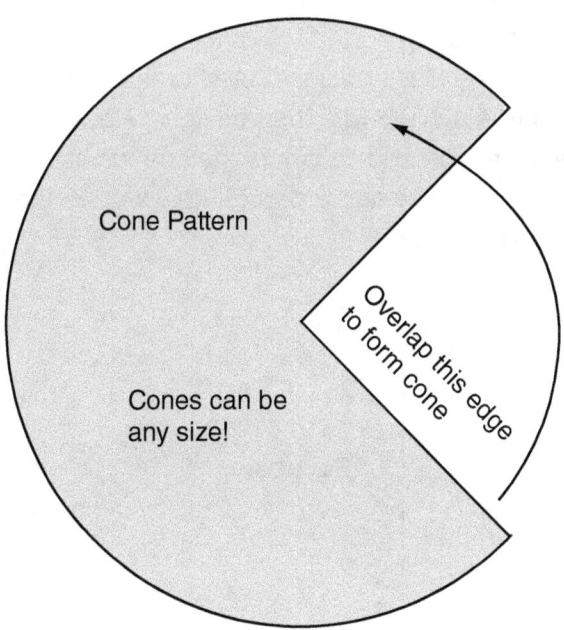

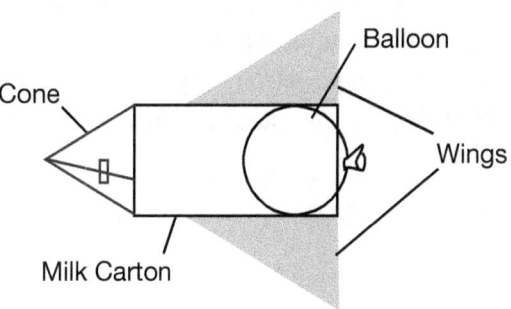

   g. Find an area at least 10 ft (3.05 m) in length for the Shuttle to travel on a smooth tile floor (a hallway or a clear area in the classroom).
   h. Place a piece of tape on the floor for the starting line.

i. Line the path you want the Shuttle to travel with books to help keep it in a straight line.
   j. Carefully blow up the balloon as full as you can, and hold the end until ready to release it.
   k. Place the Shuttle on the floor at the starting line.
   l. As soon as you release the balloon, start the stopwatch or check the watch with the second hand. Record the amount of time the Shuttle moves.
   m. Mark where the Shuttle stops and measure the distance traveled.
3. Write the following formula on the board. Speed (in/s) = Distance (in) ÷ Time (s). If using metric rulers, substitute centimeters for inches.
4. Fill in the numbers and complete the formula for calculating speed. Show the class the math, and record the speed on the board.
5. Repeat the test several times. Ask the class why you are testing the Shuttle more than once. Explain that an experiment should be tested more than one time to check or verify the results. Many variables can change the outcome. For instance, the balloon may not have been blown up to the exact size it was the first time. Brainstorm with the class for other variables or things that could change the results.
6. Explain to the class that they are going to add a parachute. Ask them what this should do. They should respond that it would slow the Shuttle down by increasing drag.
7. Add a parachute to the back of the Shuttle, and repeat the tests again. Attach a small plastic shopping bag to the back of the Shuttle with string or yarn. See the diagram below.
8. Write the information each time on the board, and calculate the speed.

Discussion/Wrap-up
- Compare the speeds for the Shuttle with and without the parachute. Have the older students share their group charts with the class.
- Ask the class if the parachute made a difference in the speed.
- Discuss how speed brakes would slow the Shuttle down.

Extensions
Have the class design speed brakes for the Shuttle model. Test their designs. Record information and calculate the speed with speed brakes and a parachute. Discuss the class findings.

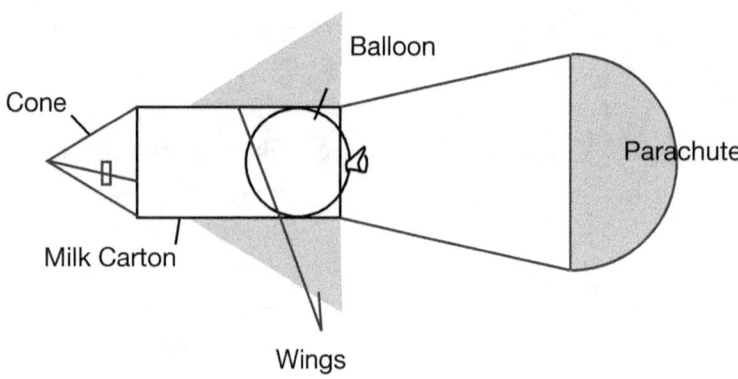

Space Shuttle Test Chart

Equipment Configuration	Parameter	Test #1	Test #2	Test #3	Average Speed (Test 1 + Test 2 + Test 3)/3
Shuttle	Distance (in)				
	Time (s)				
	Speed (in/s)				
Shuttle with Parachute	Distance (in)				
	Time (s)				
	Speed (in/s)				

(Speed (in/s) = Distance (in) ÷ Time (s))

Coming in for a Landing

Can you land the Space Shuttle like you land a jet airplane? When the Space Shuttle comes back into Earth's atmosphere, it has no power. It flies like a plane with no engine.

The Shuttle lands by gliding. It uses flight controls that let it roll, pitch and yaw. This lets the pilot land the Shuttle with more control.

To slow down the 178,000-lb (80,739.44-kg) Shuttle, the pilot uses a special brake. It is a speed brake. It increases drag to help the Shuttle slow down. You can feel drag when you stick your hand out the window while the car is moving.

When the Shuttle slows, it releases a drag parachute to help slow it down even more. Then, it rolls to a stop.

The pilot has to land right the first time. Without power, the Shuttle cannot circle around and try again. NASA likes to have a lot of room for the Shuttle to land. That is why they made the landing strips extra long.

Weather plays a large role when the Shuttle comes home. The Shuttle cannot land in rainy weather. The Shuttle cannot land in rainy weather because the coating that protects the tiles from water burns off when it launches and comes back to Earth. If water were to get under the tiles while coming back to Earth, problems could occur.

When pilots land a Space Shuttle, they face a special problem. Their bodies have not had time to adjust to gravity. They are light headed, and their bodies feel very heavy. The pilots have to work hard to land the Shuttle safely due to this sensation.

Paper Rockets

Objective
In this activity, students construct small flying rockets out of paper and propel them by blowing air through a straw.

- Target concept: Acceleration
- Preparation time: 15 min
- Duration of activity: 45 min
- Student group size: Individually or in pairs

Materials and Tools
- Scrap bond paper
- Cellophane tape
- Scissors
- Sharpened fat pencil
- Milkshake straw (slightly thinner than pencil)
- Eye protection
- Standard or metric ruler
- Masking tape or altitude trackers
- Pictures of the Sun and planets

Management
After demonstrating a completed paper rocket to the students, have them construct their own paper rockets and decorate them. See diagram on page 44.

CAUTION: Because the rockets are projectiles, make sure students wear eye protection.

When students complete the rockets, distribute straws. Select a location for flying the rockets. A room with open floor space or a hallway is preferable. Prepare the floor by marking a 32.81-ft (10-m) test range with tape measures or meter sticks laid end to end. As

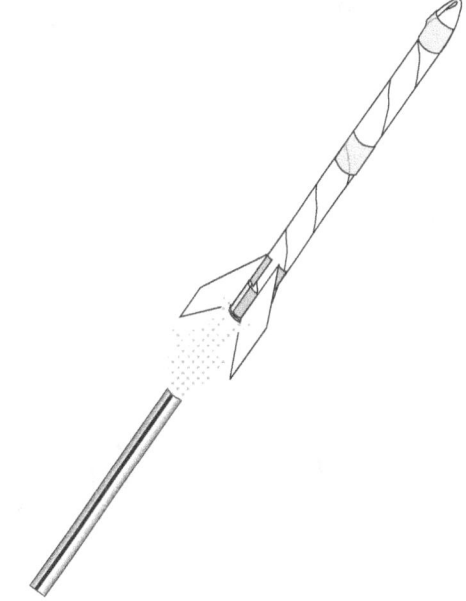

an alternative, lay out the planetary target range as shown on page 45.

Have students launch from planet Earth, and tell them to determine the farthest planet they are able to reach with their rocket. Use the planetary arrangement shown on page 45 for laying out the launch range. Pictures for the planets are found on page 46. Enlarge these pictures as desired.

Record data from each launch on the Paper Rocket Test Report form. The form includes spaces for data from three different rockets.

After the first launches, students should construct new and improved paper rockets and attempt a longer journey through the solar system. Encourage the students to try different sized rockets and different shapes and number of fins. For younger students, create a chart listing how far each planet target is from Earth. Older students can measure these distances for themselves.

How to Build a Paper Rocket

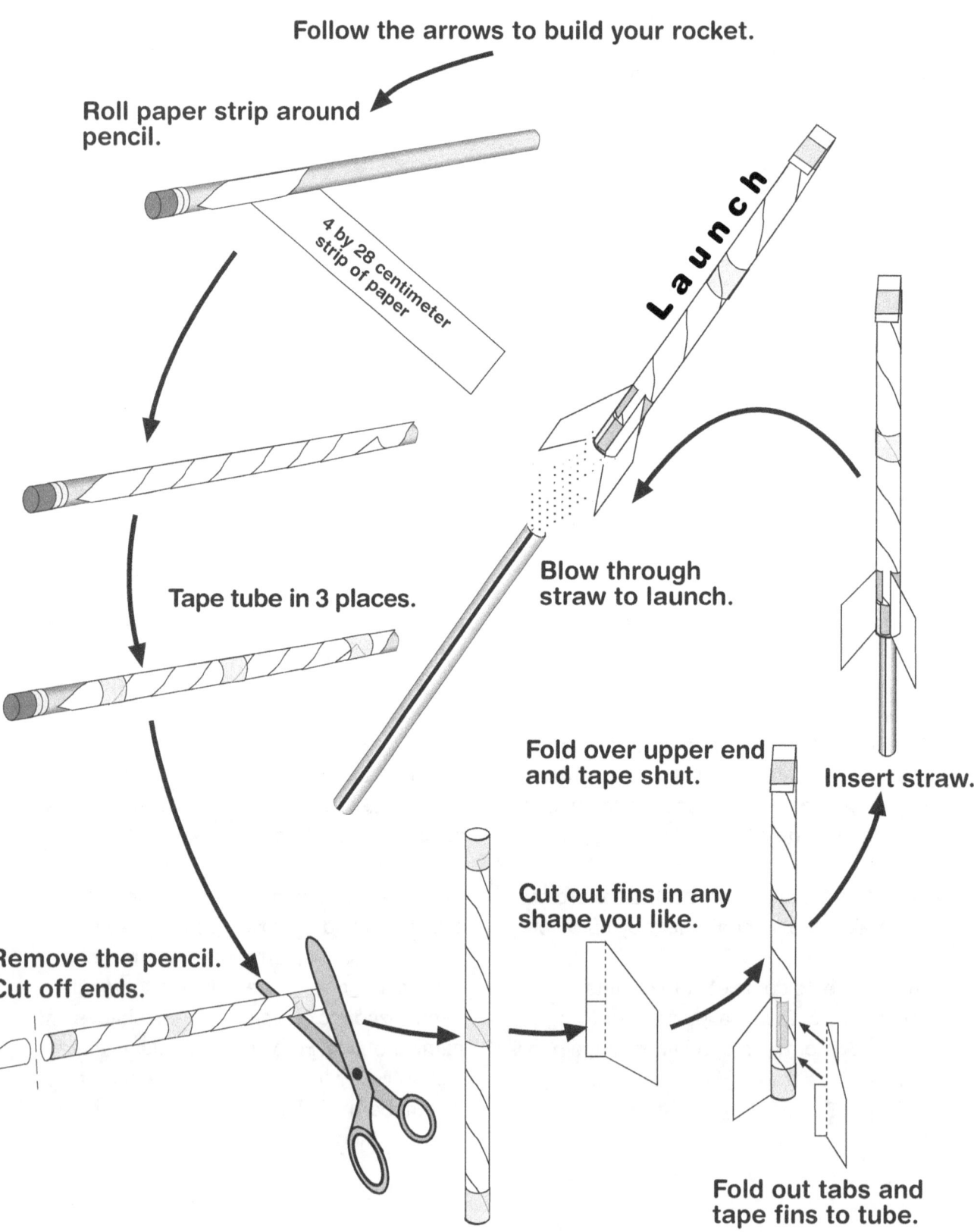

Background Information

Although the activity uses a solar system target range, the Paper Rockets activity demonstrates how rockets fly through the atmosphere. A rocket with no fins is much more difficult to control than a rocket with fins. The placement and size of the fins is critical to achieve adequate stability while not adding too much weight. More information on rocket fins can be found on page 14 of this guide.

Making and Launching Paper Rockets

1. Distribute the materials and construction tools to each student.
2. Students should each construct a rocket as shown in the instructions on the student sheet.
3. Tell students to predict how far their rocket will fly and record their estimates in the test report sheet. After test flying the rocket and measuring the distance it reached, students should record the actual distance and the difference between predicted and actual distances on the Paper Rocket Test Report form.
4. Following the flight of the first rocket, students should construct and test two additional rockets of different sizes and fin designs.

Suggested Target Range Layout

Arrange pictures of the Sun and the planets on a clear floor space as shown below. Refer to an encyclopedia or other reference for a chart on the actual distances to each planet.

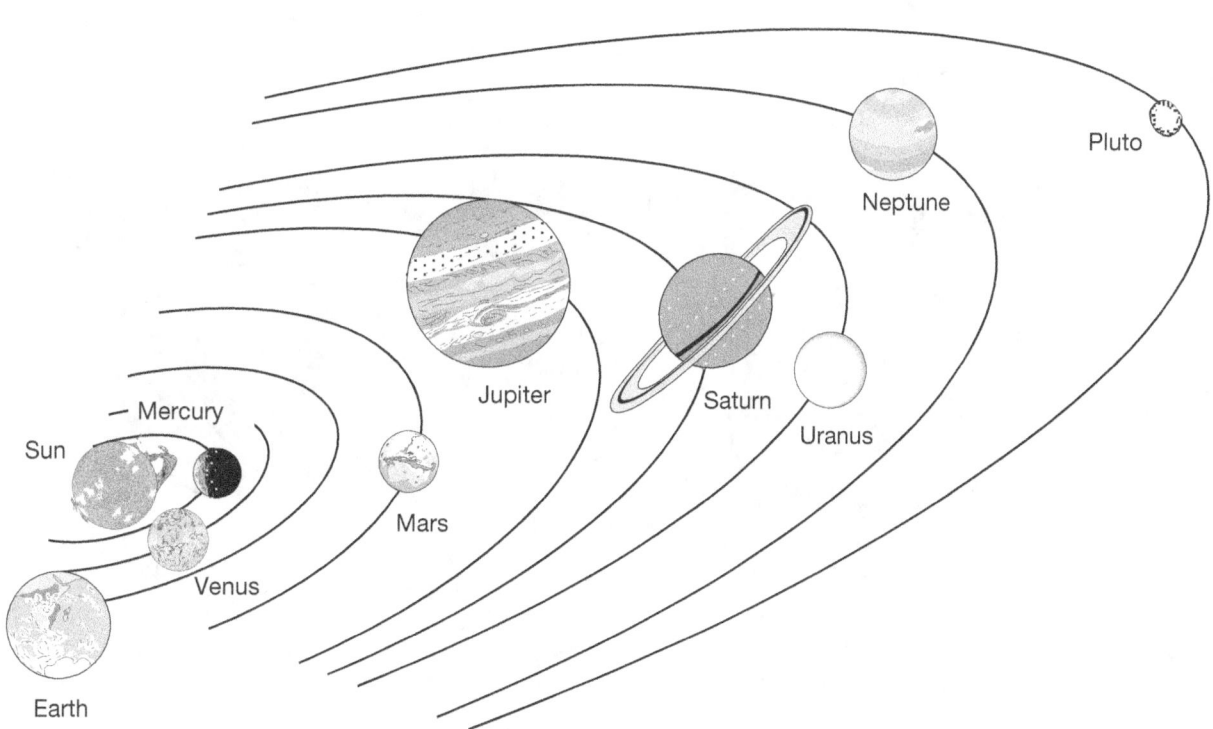

Planet Targets

(Not Drawn to Scale)

Enlarge these pictures on a copy machine or sketch copies of the pictures on separate paper. Place these pictures on the floor according to the arrangement on the previous page. If you wish to make the planets to scale, refer to the numbers beside the names indicating the relative sizes of each body. Earth's diameter is given as one and all the other bodies are given as multiples of one.

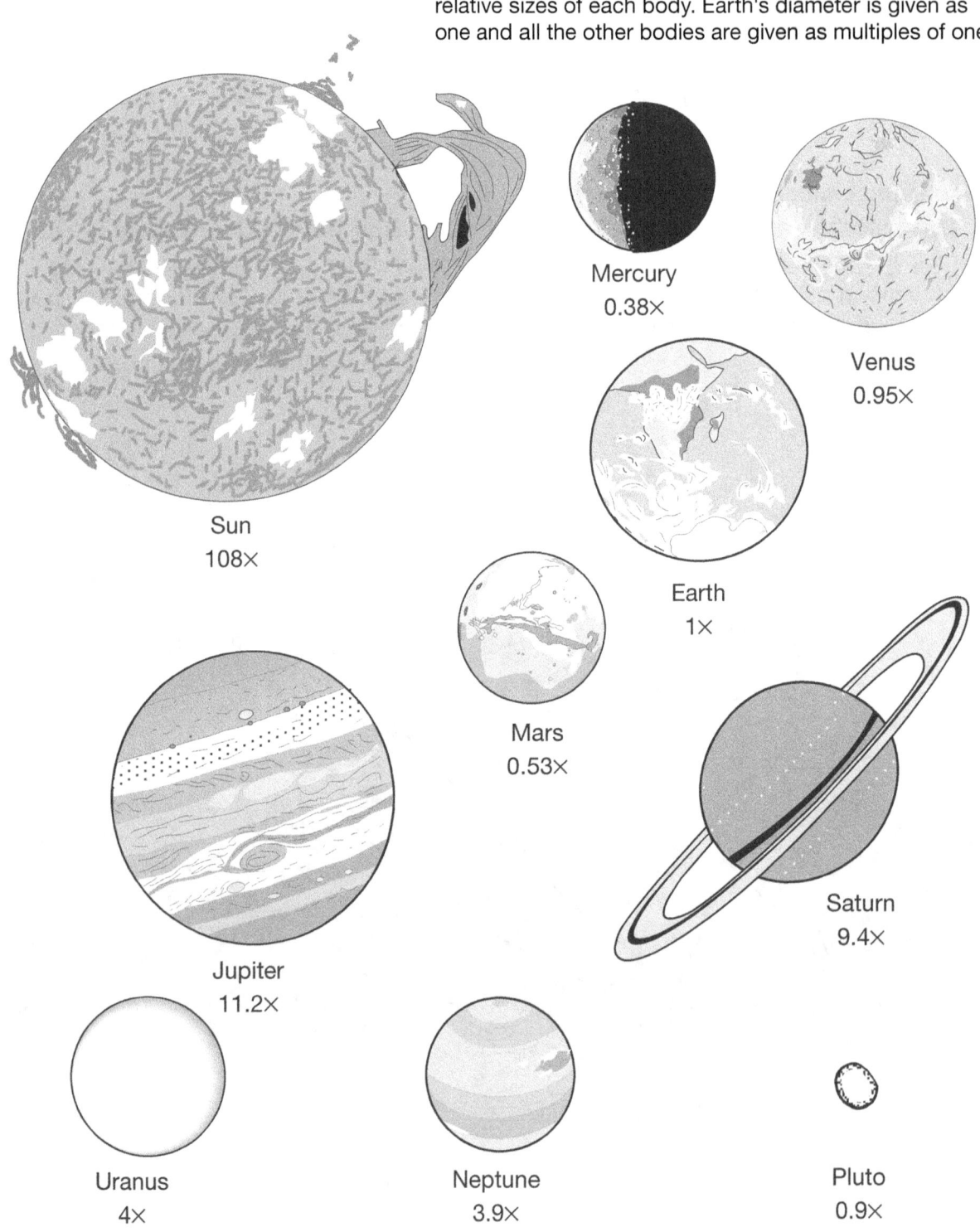

Sun 108×

Mercury 0.38×

Venus 0.95×

Earth 1×

Mars 0.53×

Jupiter 11.2×

Saturn 9.4×

Uranus 4×

Neptune 3.9×

Pluto 0.9×

Discussion

1. What makes one rocket perform better than another? (Do not forget to examine the weight of each rocket. Rockets made with extra tape and larger fins weigh more.)
2. How small can the fins be and still stabilize the rocket?
3. How many fins does a rocket need to stabilize it?
4. What would happen if you placed the rocket fins near the rocket's nose?
5. What will happen to the rocket if you bend the lower tips of the fins pinwheel fashion?
6. Are rocket fins necessary in outer space?

Extensions

Try to determine how high the rockets fly. To do so, place masking tape markers on a wall at measured distances from the floor to the ceiling. While one student launches the rocket along the wall, another student compares the height the rocket reached with the tape markers. Be sure to have the students subtract the height from where the rocket was launched from the altitude reached. For example, if students held the rocket 4.92 ft (1.5 m) from the floor to launch it, and it reached 13.12 ft (4 m) above the floor, the actual altitude change was 8.20 ft (2.5 m). Refer to the Altitude Tracker activity starting on page 109 for details on a second method for measuring the height the paper rockets reach.

Assessment

Students will complete test reports that will describe the rockets they constructed and how those rockets performed. Ask the students to create bar graphs on a blank sheet of paper that show how far each of the three rockets they constructed flew. Have students write a summarizing paragraph in which they choose which rocket performed the best and explain their ideas for why it performed as it did.

Paper Rocket Test Report

Names: _____
_____

1. Launch your rocket three times. How far did it fly each time? What is the average distance your rocket flew? Write your answer in the spaces below.
2. Build and fly a rocket of a new design. Before flying it, predict how far it will go. Fly the rocket three times and average the distances. What is the difference between your prediction and actual average distance?
3. Build a third rocket and repeat step 2.
4. On the back of this paper, write a short paragraph describing each rocket you built and how it flew. Draw pictures of the rockets you constructed.

Rocket 1 — Make notes about the flights here.

How far did it fly in inches (centimeters)?	1. _____
	2. _____
	3. _____
Average distance in inches (centimeters)? _____	

Rocket 2 — Make notes about the flights here.

Predict how many inches (centimeters) your rocket wil fly.

How far did it fly in inches (centimeters)	1. _____
	2. _____
	3. _____

Average distance? _____

Difference between your prediction and the average distance? _____

Rocket 3 — Make notes about the flights here.

Predict how many inches (centimeters) your rocket wil fly.

How far did it fly in inches (centimeters)	1. _____
	2. _____
	3. _____

Average distance? _____

Difference between your prediction and the average distance? _____

Straw Rockets

Objective
Students will be able to construct straw rockets and determine the greatest height and distance that can be reached.

- Target concept: Altitude
- Preparation time: 10 min
- Duration of activity: 40–45 min
- Student group size: Individually or in pairs

Material
- Straw rocket launcher from Appendix A
- Straws (must fit with little space remaining around the launcher rod)
- Clay
- Scissors
- Tape
- Paper
- Graph paper
- Pencils
- Protractors
- Rulers, yardstick, meter stick or measuring tapes

Assembly
Make the nose cones by molding clay into one end of the straw. On the other end of the straw tube tape on paper shapes to make fins.

Management
After demonstrating a completed straw rocket to the students, have them construct their own straw rockets and decorate them. Participant groups design and build straw rockets, launch with a preset force, measure distance and altitude, record, and check.

Straw Rocket Challenge
1. Try launch angles between 10 and 90 degrees to determine the launch angle that provides the greatest range.
2. Try launch angles between 10 and 90 degrees to determine the launch angle that provides the greatest altitude.
3. Build four rockets of different lengths. Use the same nose cone shape and fin shape. Launch the rockets at various angles and launch rod marks and determine the rocket with the most range and the rocket with the greatest altitude.
4. Build four rockets of different nose cone styles or sizes. Keep the length and fins the same to determine the effect of the nose cone shape to range and altitude.
5. Build four rockets with different numbers of fins. Keep the length and nose cone style the same to determine the effect of the number of fins on the range.
6. Using the information you gained in these experiments, design a rocket that will travel as close to 236.22 in (600 cm) as possible. You will need to determine the launch angle and launch rod settings to come as close as possible to the first launch. See how many launches it takes to achieve the 196.85 in (500 cm) range.

Rocket Racer

Objectives
- To construct a rocket propelled vehicle.
- To experiment with ways of increasing the distance the rocket racer travels.
- Target concept: Velocity
- Preparation time: 15 min
- Duration of activity: 30–40 min
- Student group size: Individually or in pairs

Materials and Tools
- Four pins
- Styrofoam meat tray
- Styrofoam cups
- Masking tape
- Flexible straw
- Scissors
- Drawing compass*
- Marker pen
- Small round party balloon
- Ruler
- Student Sheets (one set per group)
- 32.81 ft (10 m) tape measure or other measuring markers for track (one for the whole class)
- Graph paper

*If compasses are not available, students can trace circular objects to make the wheels or use the wheel and hubcap patterns printed on page 53.

Note: Styrofoam cup bottoms can also be used as wheels. Putting hubcaps on both sides of the wheels may improve performance.

Management
Students will construct a balloon-powered rocket racer from a styrofoam tray, pins, tape, and a flexible straw, and test it along a measured track on the floor. The activity stresses technology education and provides students with the opportunity to modify their racer designs to increase performance. The optional second part of the activity directs students to design, construct and test a new rocket racer based on the results of the first racer. Refer to the materials list, and provide what is needed for making one rocket racer for each group of two students. Styrofoam food trays are available from butchers in supermarkets. They are usually sold for a few cents each or you may be able to get them donated. Students can also save trays at home and bring them to class.

If using the second part of the activity, provide each group with an extra set of materials. Save scraps from the first styrofoam tray to build the second racer. You may wish to hold drag or distance races with the racers. The racers will work very well on tile floors and carpeted floors with a short nap. Several tables stretched end to end will also work, but racers may roll off the edges.

Although this activity provides one rocket racer design, students can try any racer shape and any number, size and placement of wheels they wish. Long racers often work differently than short racers.

Background Information
The Rocket Racer is a simple way to observe Newton's Third Law of Motion. (Refer to page 162 of the rocket principles section of this guide for a complete description.) While it is possible to demonstrate Newton's Third Law with just a balloon (attached to fishing line by a straw), constructing a rocket racer provides students with the opportunity to put the action/reaction force to practical use. In this case, the payload of the balloon rocket is the racer. Wheels reduce friction with the floor to help racers move. Because of individual variations in the student racers, they will travel different distances and many times in unplanned directions. Through modifications, the students can correct for undesirable results and improve their racers' efficiency.

Extensions
Hold Rocket Racer races.

- Tie a loop of string around the inflated balloon before releasing the racer. Inflate the balloon inside the string loop each time you test the racers. This will increase the accuracy of the tests by ensuring the balloon inflates the same amount each time.
- Make a balloon-powered pinwheel by taping another balloon to a flexible straw. Push a pin through the straw and into the eraser of a pencil. Inflate the balloon and watch it go.

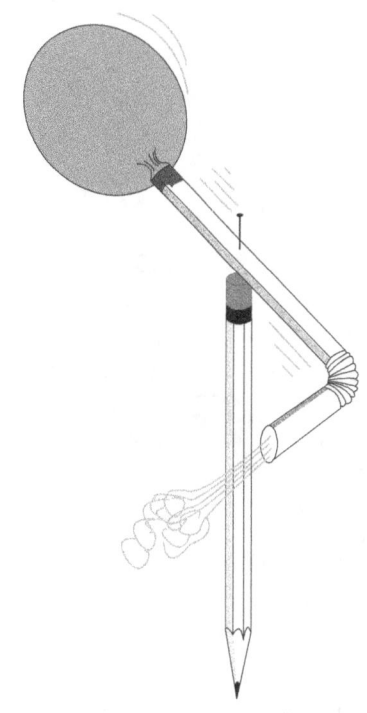

How to Build a Rocket Racer

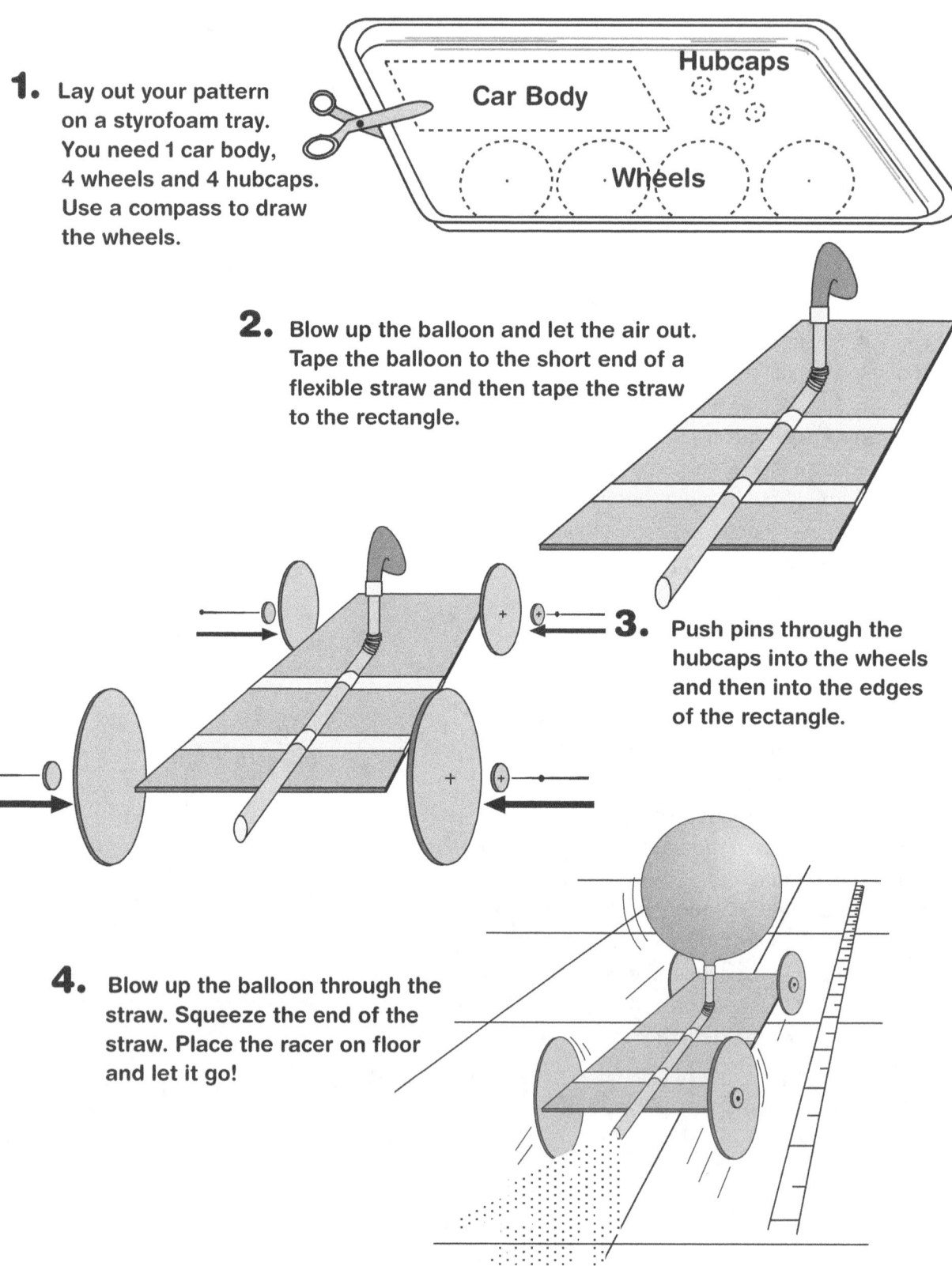

1. Lay out your pattern on a styrofoam tray. You need 1 car body, 4 wheels and 4 hubcaps. Use a compass to draw the wheels.

2. Blow up the balloon and let the air out. Tape the balloon to the short end of a flexible straw and then tape the straw to the rectangle.

3. Push pins through the hubcaps into the wheels and then into the edges of the rectangle.

4. Blow up the balloon through the straw. Squeeze the end of the straw. Place the racer on floor and let it go!

Wheel Patterns
(Crosses mark the centers.)

Hubcap Patterns
(Crosses mark the centers.)

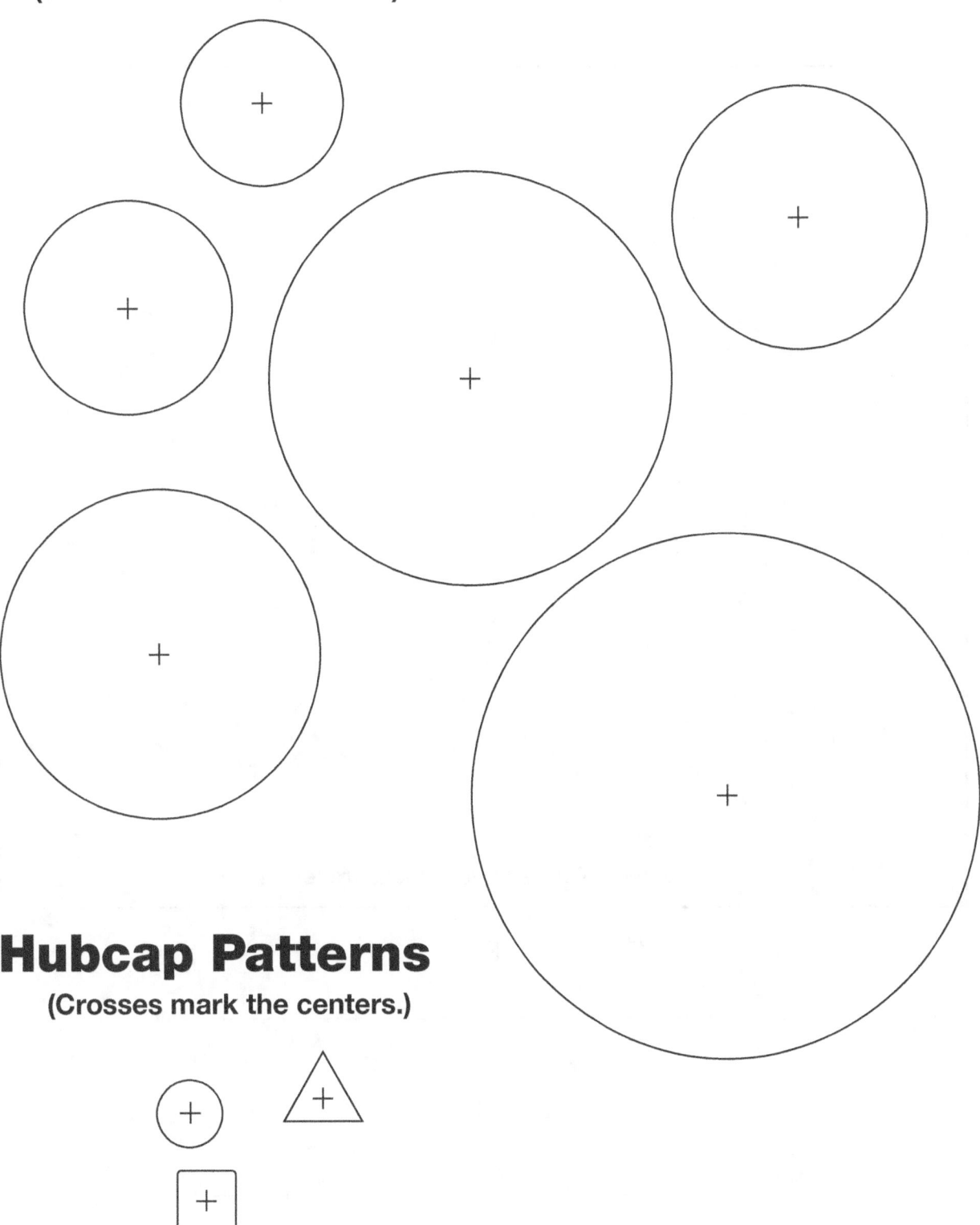

53

Rocket Racer Test Report

Draw a picture of your rocket racer.

By

_____

_____

Date: _____

Rocket Racer Test Report

Place your rocket racer on the test track and measure hw far it travels.

1. Describe how your rocket racer ran during the first trial run.
   (Did it run on a straight or curved path?)
   _____

   How far did it go? _____ inches (centimeters).

   Color in one block on the graph for each 3.94 in (10 cm) your racer traveled.

2. Find a way to change and improve your rocket racer and test it again.
   What did you do to improve the rocket racer for the second trial run?

   _____

   _____

   How far did it go?_____ inches (centimeters)_____
   Color in one block on the graph for each 3.94 in (10 cm) your racer traveled.

3. Find a way to change and imporve your rocket racer and test it again.
   What did you do to improve the rocket racer for the third trial run?

   _____

   _____

   How far did it go?_____ inches (centimeters)_____
   Color in one block on the graph for each 3.94 in (10 cm) your racer traveled.

4. In which test did your racer go the farthest? _____

   Why?_____

   _____

Rocket Racer Data Sheet

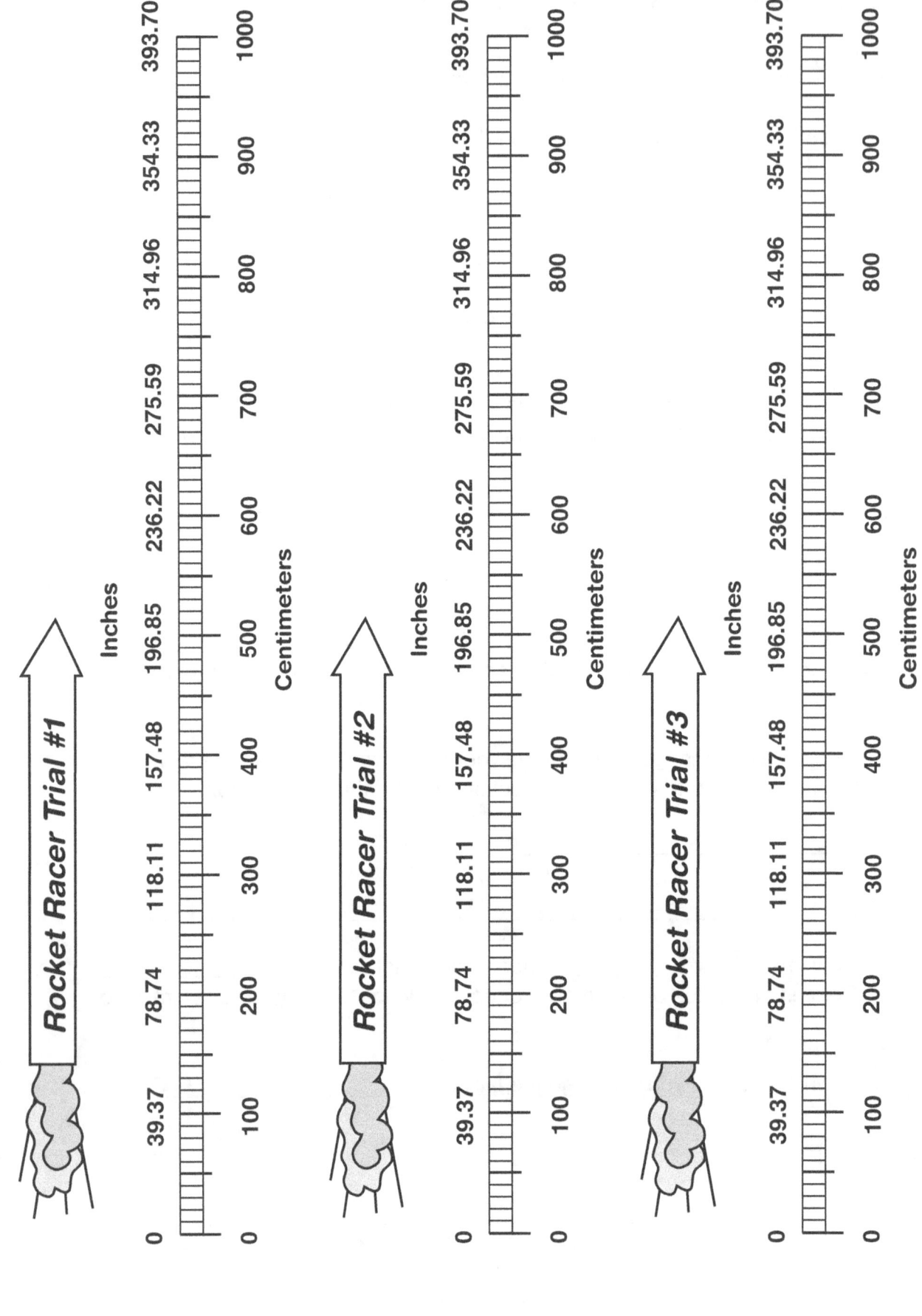

56

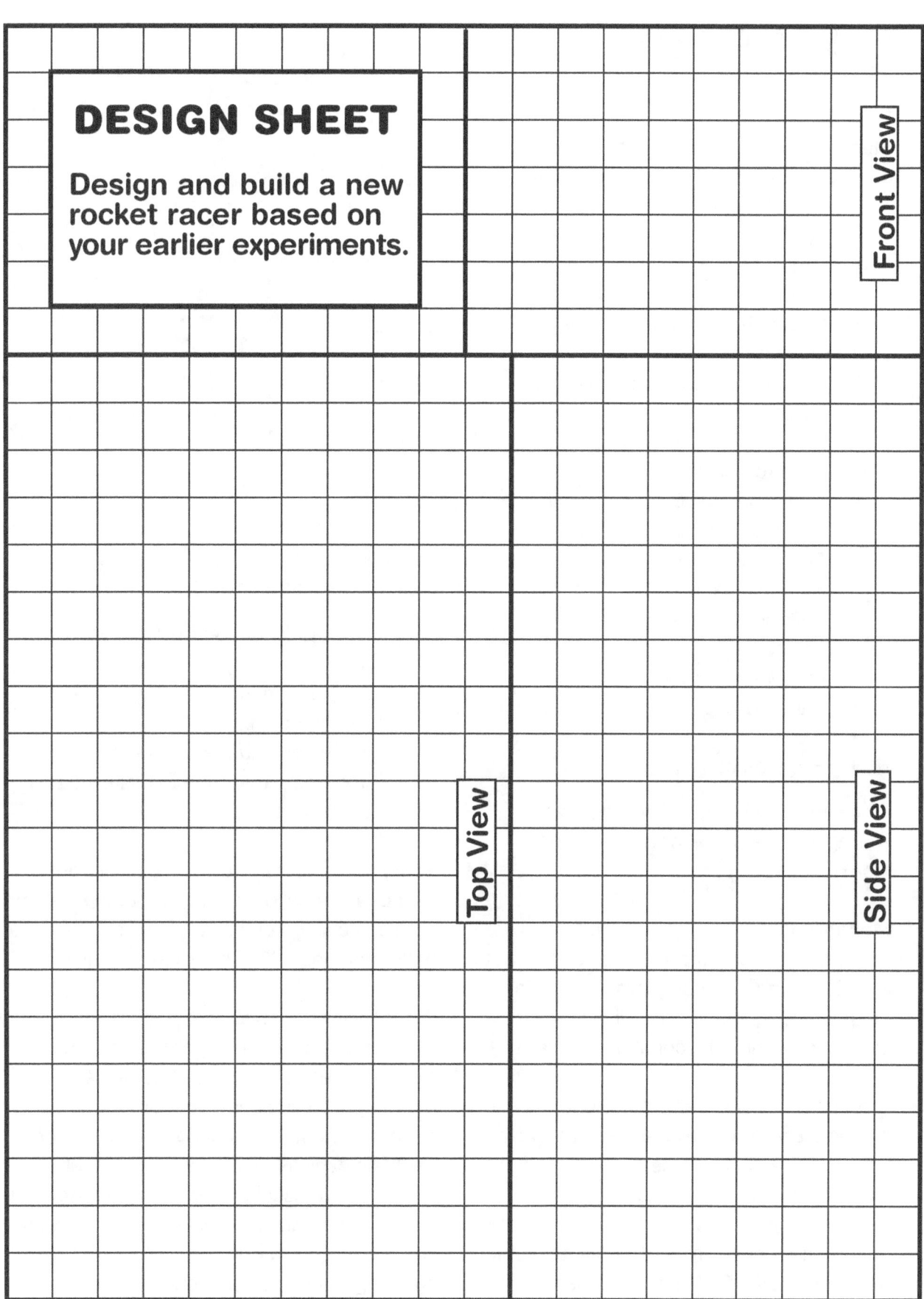

3-2-1 POP!

Objective
Students construct a rocket powered by the pressure generated from an effervescing antacid tablet reacting with water.

- Target concept: Altitude, velocity and acceleration
- Preparation time: 10 minutes
- Duration of activity: 40–45 minutes
- Student group size: Teams of two students

Materials and Tools
- Heavy paper (60–110 index stock or construction paper)
- Plastic 35 mm film canister*
- Student sheets
- Cellophane tape
- Scissors
- Effervescing antacid tablet
- Paper towels
- Water
- Eye protection

*The film canister must have an internal-sealing lid. See management section for more details.

Management
It may be helpful to make samples of rockets in various stages of completion available for students to study. This will help some students visualize the construction steps.

A single sheet of paper is sufficient to make a rocket. Be sure to tell the students to plan how they are going to use the paper. Let the students decide whether to cut the paper the short or long direction to make the body tube of the rocket. This will lead to rockets of different lengths for flight comparison.

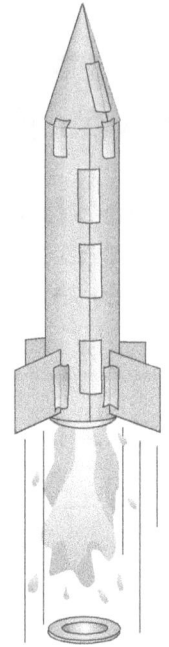

The most common mistakes in constructing the rocket are as follows:

- Forgetting to tape the film canister to the rocket body.
- Failing to mount the canister with the lid end down.
- Not extending the canister far enough from the paper tube to make snapping the lid easy.

Some students may have difficulty in forming the cone. To make a cone, cut out a pie shape from a circle and curl it into a cone. See the pattern on page 60. Cones can be any size.

Film canisters are available from camera shops and stores where photographic processing takes place. These businesses recycle the canisters and are often willing to donate them for educational use. Be sure to obtain canisters with the internal sealing lid. These are usually translucent canisters. Canisters with the external lid (lid that wraps around the canister rim) will not work. These are usually opaque canisters.

Background Information

This activity is a simple, but exciting, demonstration of Newton's Laws of Motion. The rocket lifts off because an unbalanced force acts upon it (First Law). This is the force produced when the lid blows off by the gas formed in the canister. The amount of force is directly proportional to the mass of water and gas expelled from the canister and how fast it accelerates (Second Law).The rocket travels upward with a force that is equal and opposite to the downward force propelling the water, gas and lid (Third Law). For a more complete discussion of Newton's Laws of Motion, see page 159 in this guide.

Procedure

Refer to the Student Sheet on page 61.

Discussion

- How does the amount of water placed in the cylinder affect how high the rocket will fly?
- How does the temperature of the water affect how high the rocket will fly?
- How does the amount of the tablet used affect how high the rocket will fly?
- How does the length or empty weight of the rocket affect how high the rocket will fly?
- How would it be possible to create a two-stage rocket?

Assessment

Ask students to explain how Newton's Laws of Motion apply to this rocket. Compare the rockets for skill in construction. Rockets that use excessive paper and tape are likely to be less efficient fliers because they carry additional weight.

Extensions

- Hold an altitude contest to see which rockets fly the highest. Launch the rockets near a wall in a room with a high ceiling. Tape a tape measure to the wall. Stand back and observe how high the rockets travel upward along the wall. Let all students take turns measuring rocket altitudes.
- What geometric shapes are present in a rocket?
- Use the discussion questions to design experiments with the rockets. Graph your results.

3-2-1 POP!

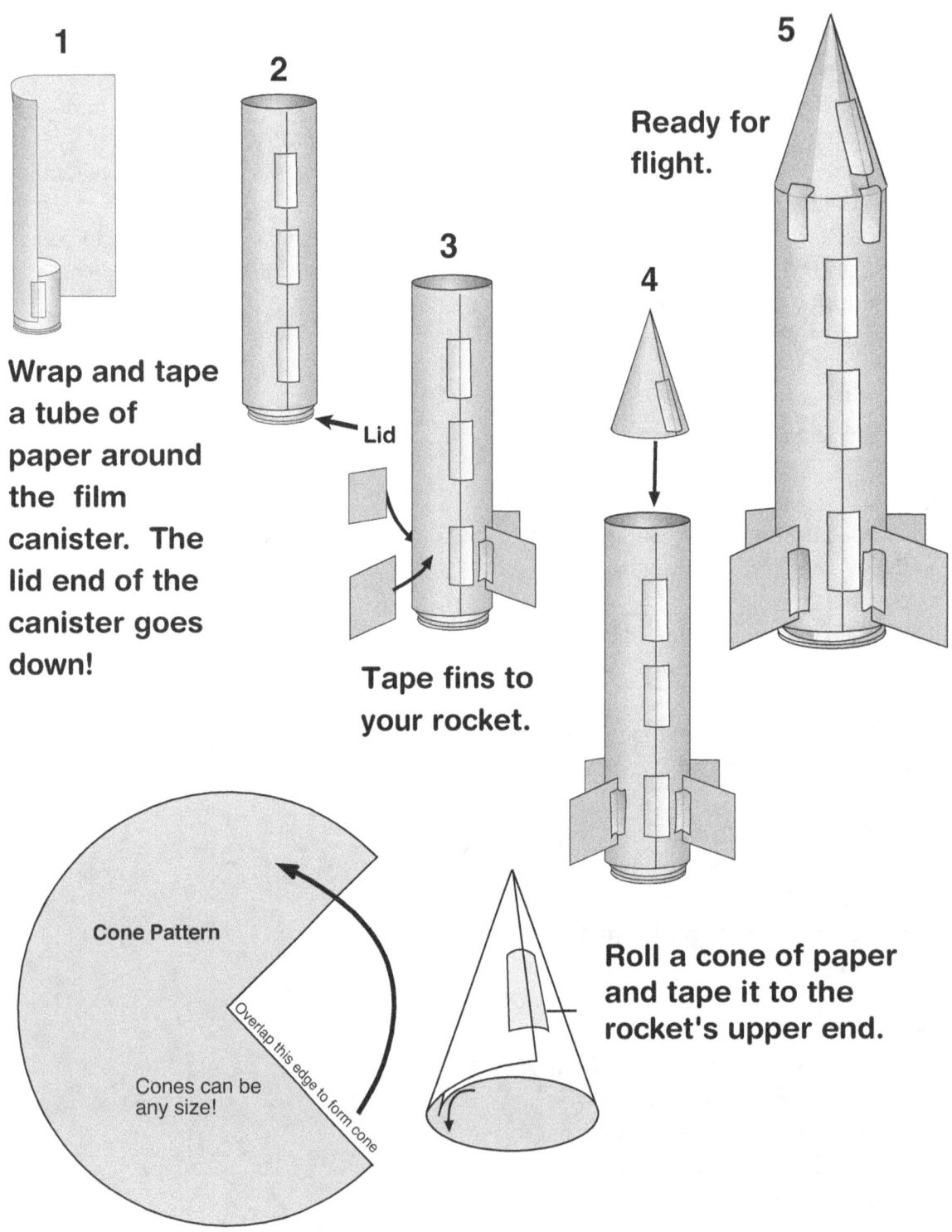

1. Wrap and tape a tube of paper around the film canister. The lid end of the canister goes down!

2. Lid

3. Tape fins to your rocket.

4. Roll a cone of paper and tape it to the rocket's upper end.

5. Ready for flight.

Cone Pattern

Cones can be any size!

Overlap this edge to form cone

60

Rocketeer Names

_____

_____

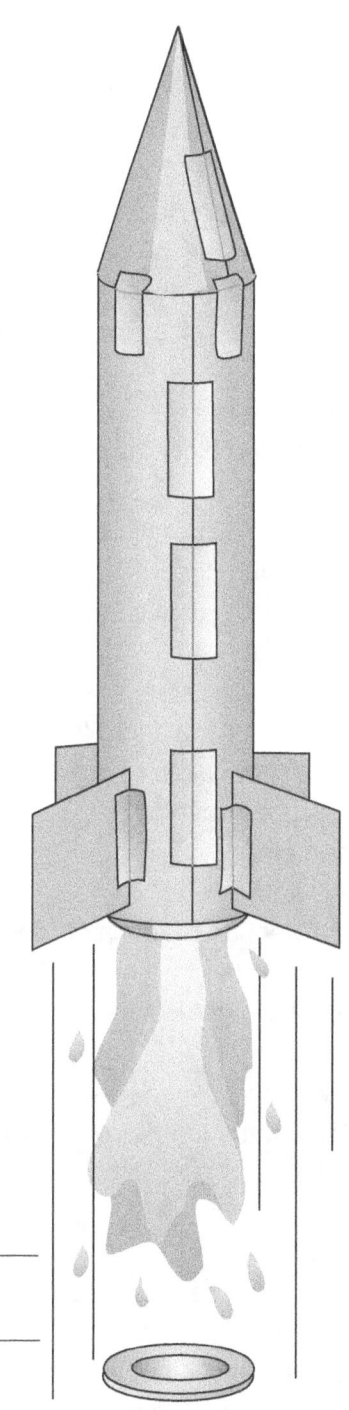

COUNTDOWN:

1. Put on your eye protection.
2. Turn the rocket upside down and fill the canister one-third full of water.

Work quickly on the next steps!

3. Drop in 1/2 tablet.
4. Snap lid on tight.
5. Stand rocket on launch platform.
6. Stand back.

LIFT-OFF!

What three ways can you improve your rocket?

1. _____

2. _____

3. _____

Newton Car

Objective
In this activity, students will test a slingshot-like device that throws a mass causing the car to move in the opposite direction.

- Target concept: Acceleration
- Preparation time: More than 30 min
- Duration of activity: 45–60 min
- Student group size: Teams of three

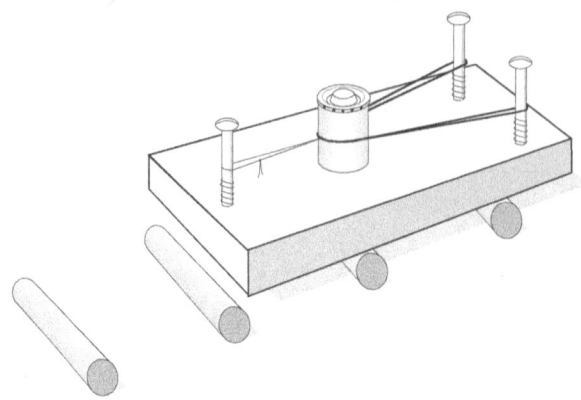

Materials and Tools
- One wooden block about 3.94 × 7.87 × 0.98 in (10 × 20 × 2.5 cm)
- Three 3-in (7.62 cm) No. 10 wood screws (round head)
- Twelve round pencils or short lengths of similar dowel
- Plastic film canister
- Assorted materials for filling canister (washers, nuts, etc.)
- Three rubber bands
- Cotton string
- Safety lighter
- Eye protection for each student
- American Standard or metric beam balance (Primer Balance)
- Vise
- Screwdriver
- Yardstick, meter stick or rulers
- Graph paper

Management
Use a smooth testing surface such as a long, level tabletop or uncarpeted floor. The experiment has many variables that students must control including the size of the string loop they tie, the placement of the mass on the car and the placement of the dowels. Discuss the importance of controlling the variables in the experiment with your students.

Making the Newton Car involves cutting blocks of wood and driving three screws into each block. Refer to the diagram on this page for the placement of the screws as well as how the Newton Car is set up for the experiment. Place the dowels in a row like railroad ties and extend them to one side as shown in the picture. If you have access to a drill press, you can substitute short dowels for the screws. It is important to drill the holes for the dowels perpendicular into the block with the drill press. Add a drop of glue to each hole.

The activity requires students to load their slingshot by stretching the rubber bands back to the third screw and holding them in place with the string. The simplest way of doing this is to tie the loop first and slide the rubber bands through the loop before placing the rubber bands over the two screws. Loop the string over the third screw after stretching the rubber bands back.

Use a match or lighter to burn the string. The small ends of string left over from the knot acts as a fuse that permits the students to remove the match before the string burns through. Teachers may want to give student groups only a few matches at a time.

To completely conduct this experiment, student groups will need six matches. It may be necessary for a practice run before starting the experiment. As an alternative to the matches, students can use blunt nose scissors to cut the string. This requires some fast movement on the part of the student doing the cutting. The student needs to move the scissors quickly out of the way after cutting the string.

Tell the students to tie all the string loops they need before beginning the experiment. The loops should be as close to the same size as possible. Refer to the diagram on the student pages for the actual size of the loops. Loops of different sizes will introduce a significant variable into the experiment, causing the rubber bands to be stretched different amounts. This will lead to different accelerations with the mass each time the experiment is conducted.

Use plastic 35-millimeter film canisters for the mass in the experiments. Direct the students to completely fill the canisters with various materials such as seeds, small nails, metal washers, sand, etc. This will enable them to vary the mass twice during the experiment. Have the students weigh the canister after it is filled and record the mass on the student sheet. After using the canister three times, first with one rubber band and then two and three rubber bands, students should refill the canister with new material for the next three tests.

Refer to the sample graph for recording data. The bottom of the graph is the distance the car travels in each test. Students should plot a dot on the graph for the distance the car traveled. The dot should fall on the y-axis line representing the number of rubber bands used and on the x-axis for the distance the car traveled. After plotting three tests with a particular mass, connect the dots with lines. The students should use a solid line for Mass 1 and a line with large dashes for Mass 2. If the students have carefully controlled their variables, they should observe that the car traveled the greatest distance using the greatest mass and three rubber bands. This conclusion will help them conceptualize Newton's Second Law of Motion.

Background Information

The Newton Car provides an excellent tool for investigating Newton's Second Law of Motion, which states that force equals mass times acceleration. In rockets, the force is the action produced by gas expelled from the engines. According to the law, the greater the mass of gas that is expelled and the faster it accelerates out of the engine, the greater the force or thrust. More details on this law can be found on page 163 of this guide.

The Newton Car is a kind of a slingshot. A wooden block with three screws driven into it forms the slingshot frame. Rubber bands stretch from two of the screws and hold to the third by a string loop. A mass sits between the rubber bands. When the string is cut, the rubber bands throw the block to produce an action force. The reaction force propels the block in the opposite direction over some dowels that act as rollers (Newton's Third Law of Motion).

This experiment directs students to launch the car while varying the number of rubber bands and the quantity of mass thrown off. They will measure how far the car travels in the opposite direction and plot the data on a graph. Repeated runs of the experiment should show that the distance the car travels depends on the number of rubber bands used and the quantity of the mass being expelled. Comparing the graph lines will lead students to Newton's Second Law of Motion.

Discussion
1. How is the Newton Car similar to rockets?
2. How do rocket engines increase their thrust?
3. Why is it important to control variables in an experiment?

Assessment:
Conduct a class discussion where students share their findings about Newton's Laws of Motion. Ask them to compare their results with those from other activities, such as Pop Can Hero Engine. Collect and review completed student pages.

Extensions
Obtain a toy water rocket from a toy store. Try launching the rocket with only air and then with water and air and observe how far the rocket travels.

Newton Car

1. Tie six string loops this size.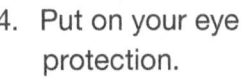

2. Fill up your canister and weigh it in ounces (grams). Record the weight (mass) in the Newton Car Report.

3. Set up your Newton Car as shown in the picture. Slip the rubber band through the string loop. Stretch the rubber band over the two screws and pull the string back over the third screw. Place the rods 2.5 in (6.35 cm) apart. Use only one rubber band the first time.

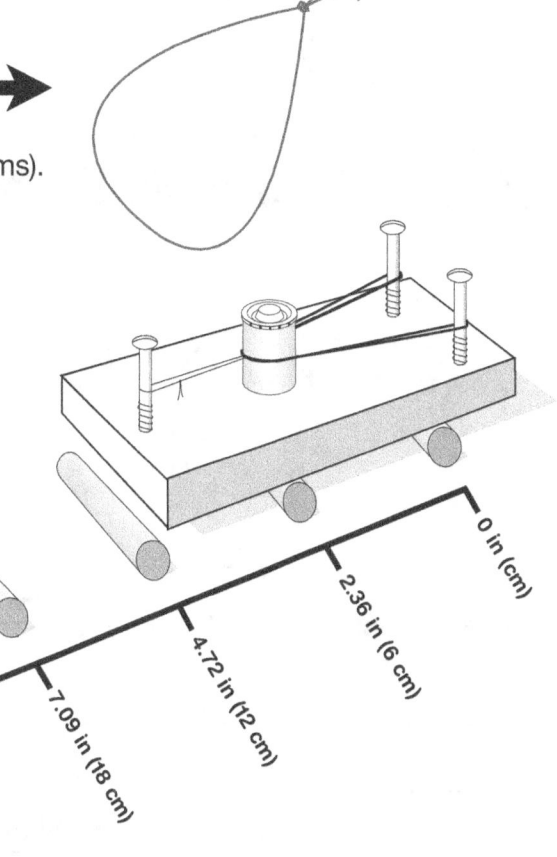

4. Put on your eye protection.

5. Light the string with a match and stand back. Record the distance the car traveled on the chart.

6. Reset the car and rods. Make sure that the rods are 2.5 in (6.35 cm) apart. Use two rubber bands and record the distance the car travels.

7. Reset the car with three rubber bands and repeat the experiment. Record the distance it travels.

8. Refill the canister, and record its new mass.

9. Test the car with the new canister and with one, two and three rubber bands. Record the distances the car moves each time.

10. Plot your results on the graph. Use one line for the first set of measurements and a different line for the second set.

Newton Car Report

Team _____

Members _____

_____

MASS 1

____ oz (gms)

Rubber Bands	Distance Traveled
	_____ in (cm)
	_____ in (cm)
	_____ in (cm)

Describe what happened when you tested the car with one, two and three rubber bands.

_____

_____

MASS 2

____ oz (gms)

Rubber Bands	Distance Traveled
	_____ in (cm)
	_____ in (cm)
	_____ in (cm)

Describe what happened when you tested the car with one, two and three rubber bands.

_____

_____

Write a short statement explaining the relationship between the amount of mass in the canister, the number of rubber bands and the distance the car traveled.

_____

_____

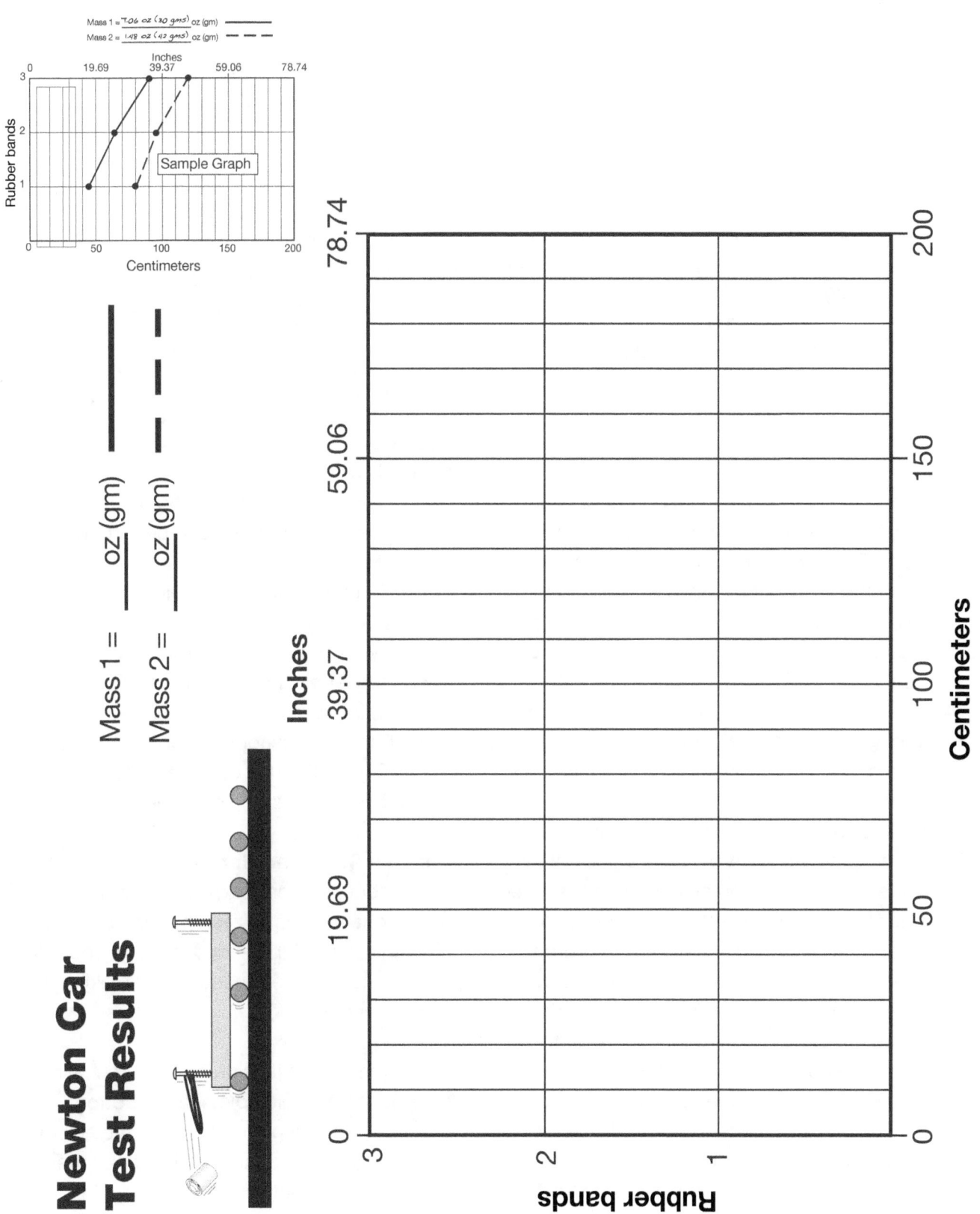

Pop Can Hero Engine

Objectives
- Part One: To demonstrate action and reactions by using the force of falling water to cause a soda pop can to spin.
- Part Two: To experiment with different ways of increasing the spin of the can.
- Target concept: Velocity
- Preparation time: Variable
- Duration of activity: 45 min
- Student group size: Teams of two or three

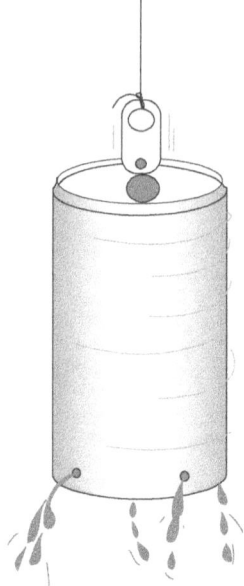

Management
This activity is divided into two parts. In part one, the learners construct the engine and test it. Part two focuses on variables that affect the action of the engine. The experiment stresses prediction, data collection and analysis of results. Be sure to recycle the soda cans at the end of the activity.

Background Information
Hero of Alexandria invented the Hero engine in the first century B.C. His engine operated because of the propulsive force generated by escaping steam. A boiler produced steam that escaped to the outside through L-shaped tubes bent pinwheel fashion. The steam's escape produced an action-reaction force that caused the sphere to spin in the opposite direction. Hero's engine is an excellent demonstration of Newton's Third Law of Motion (See page 18 for more information about Hero's engine and page 162 for details about Newton's Third Law of Motion.). This activity substitutes the action force produced by falling water for the steam in Hero's engine.

Part One
Materials and Tools
- Empty soda can with the opener lever still attached (one per group of students)
- Common nail (one per group of students)
- Nylon fishing line (lightweight)
- Bucket or tub of water (several for entire class)
- Paper towels for cleanup
- Yard stick, ruler or meter stick
- Scissors to cut fishing line

Making a Soda Can Hero Engine
1. Distribute student pages, one soda can and one medium-size common nail to each group. Tell the students that you will demonstrate the procedure for making the Hero engine.

2. Lay the can on its side and use the nail to punch a single hole near its bottom. Before removing the nail, push the nail to one side to bend the metal, making the hole slant in that direction.

How To Bend The Holes

Punch hole with nail.

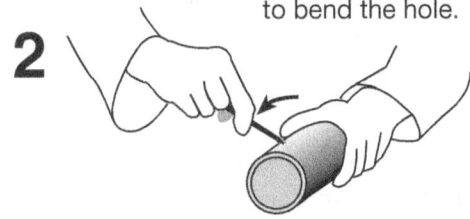

With nail still inserted, push upper end of nail to the side to bend the hole.

3. Remove the nail and rotate the can approximately 90 degrees. Make a second hole like the first one. Repeat this procedure two more times to produce four equally spaced holes around the bottom of the can. All four holes should slant in the same direction going around the can.

4. Bend the can's opener lever straight up and tie a 15–20-in (38.1–50.8-cm) length of fishing line to it. The soda can Hero engine is complete.

Running the Engine

1. Dip the can in the water tub until it fills with water. Ask the students to predict what will happen when you pull the can out by the fishing line.
2. Have each group try out their Hero engine.

Discussion

1. Why did the cans begin spinning when water poured out of the holes?
2. What was the action? What was the reaction?
3. Did all cans spin equally well? Why or why not?

Part Two
Materials and Tools

- Student Work Sheets
- Hero engines from part one
- Empty soda can with the opener lever still attached (three per group of students)
- Common nails: two different diameter shafts (one each per group)
- Nylon fishing line (lightweight)
- Bucket or tub of water (several for entire class)
- Paper towels for cleanup
- Yard stick, rulers or meter stick
- Large round colored gum labels or marker pens
- Scissors to cut fishing line

Experimenting With Soda Can Hero Engines

1. Tell the students they are going to do an experiment to find out if there is any relationship between the size of the holes punched in the Hero engine and how many times it rotates. Ask students to predict what they think might happen to the rotation of the Hero engine if they punched larger or smaller holes in the cans. Discuss possible hypotheses for the experiment.

2. Provide each group with the materials listed for Part Two. The nails should have different diameter shafts from the one used to make the first engine. Identify these nails as small (S) and large (L). Older students can measure the diameters of the holes in inches (millimeters). Since there will be individual variations, record the average hole diameter. Have the groups make two additional engines exactly like the first, except that the holes will be different sizes.

3. Discuss how to count the times the engines rotate. To aid in counting the number of rotations, stick a brightly colored round gum label or some other marker on the can. Tell them to practice counting the rotations of the cans several times to become consistent in their measurements before running the actual experiment.
4. Have the students write their answers for each of three tests they will conduct on the can diagrams on the Student pages. (Test one employs the can created in Part One.) Students should not predict results for the second and third cans until they have finished the previous tests.
5. Discuss the results of each group's experiment. Did the results confirm the experiment hypothesis?
6. Ask the students to propose other ways of changing the engine's rotation (make holes at different distances above the bottom of the can, slant holes in different directions or not slanted at all, etc.). Be sure they compare the fourth Hero engine they make with the engine previously made that has the same size holes.

Discussion
1. Compare the way rockets in space change the directions they are facing in space with the way Hero engines work.
2. How can you get a Hero engine to turn in the opposite direction?
3. Can you think of any way to put Hero engines to practical use?
4. In what ways are Hero engines similar to rockets? In what ways are they different?

Assessment
Conduct a discussion where students share their findings about Newton's Laws of Motion. Collect and review completed Student pages.

Extensions
- Compare a rotary lawn sprinkler to Hero's engine.
- Research Hero and his engine. Was the engine put to any use?
- Build a steam-powered Hero engine (See instructions below).

Teacher Model
Steam-Powered Hero Engine (Demonstration Model)
This should be created and operated by an adult. A steam powered Hero engine can be manufactured from a copper toilet tank float and some copper tubing. Because this version of the Hero engine involves steam, it is best to use it as a demonstration only.

Materials and Tools
- Copper toilet tank float (available from some hardware or plumbing supply stores)
- Thumbscrew: 0.25 in (0.635 cm)
- Brass tube: 0.1875-in (0.476-cm) inner diameter and 12-in (30.18-cm) long (from hobby shops)
- Solder
- Fishing line
- Ice pick or drill
- Metal file
- Propane torch

Procedure

1. File the middle of the brass tube to produce a notch. Do not file the tube in half.

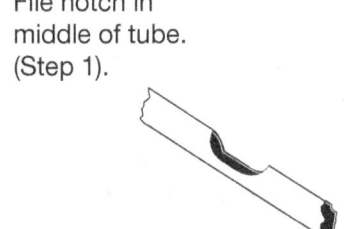

File notch in middle of tube. (Step 1).

2. Using the ice pick or drill, bore two small holes on opposite sides of the float at its middle. The holes should be just large enough to pass the tube straight through the float.
3. With the tube positioned so that equal lengths protrude through the float, heat the contact points of the float and tube with the propane torch. Touch the end of the solder to the heated area so that it melts and seals both joints.
4. Drill a water access hole through the threaded connector at the top of the float.
5. Using the torch again, heat the protruding tubes about 1.5 in (3.81 cm) from each end. With pliers, carefully bend the tube tips in opposite directions. Bend the tubes slowly so they do not crimp.
6. Drill a small hole through the flat part of the thumbscrew for attaching the fishing line and swivel. Twist the thumbscrew into the threaded connector of the float in step 4 and attach the fishing line and swivel.

Finished Steam-Powered Hero Engine

1. Place a small amount of water (about 0.5–1 oz [114.79–29.7 ml]) into the float. The precise amount is not important. The float can be filled through the top if you drilled an access hole or through the tubes by partially immersing the engine in a bowl of water with one tube submerged and the other out of the water.
2. Suspend the engine and heat its bottom with the torch. In a minute or two, the engine should begin spinning.

Be careful not to operate the engine too long because it may not be balanced well and could wobble violently. If it begins to wobble, remove the heat.

CAUTION: Wear eye protection when demonstrating the engine. Be sure to confirm that the tubes are not obstructed in any way before heating. Test them by blowing through one like a straw. If air flows out of the other tube, the engine is safe to use.

Finished Steam-Powered Hero Engine

Pop Can Hero Engine

Names of Team Members

_____   _____

_____

Design an experiment that will test the effect that the size of the holes has on the number of spins the can makes. What is your experiment hypothesis?

_____

_____

Mark each can to help you count the spins. Test each Hero engine and record your data on the can diagrams below.

Test Number 1

Hero Engine
Number of Holes: _____
Size of Holes: _____
Predicted Number of Spins: _____
Actual Number of Spins: _____
Difference (+ or -): _____

Test Number 2

Hero Engine
Number of Holes: _____
Size of Holes: _____
Predicted Number of Spins: _____
Actual Number of Spins: _____
Difference (+ or -): _____

Test Number 3

Hero Engine
Number of Holes: _____
Size of Holes: _____
Predicted Number of Spins: _____
Actual Number of Spins: _____
Difference (+ or -): _____

Based on your results, was your hypothesis correct? _____

Why? _____

_____

Design a new Hero engine experiment. Remember, change only one variable in your experiment.

What is your experiment hypothesis? _____

_____

_____

Compare this engine with the engine from your first experiment that has the same size holes.

Hero Engine

Number of Holes: _____

Size of Holes: _____

Predicted Number of Spins: _____

Actual Number of Spins: _____

Difference (+ or -): _____

Based on your results, was your hypothesis correct? _____

Why? _____

_____

_____

Describe what you learned about Newton's Laws of Motion by building and testing your Hero engines.

_____

_____

_____

Share your findings with other members of your class.

73

Rocket Transportation

Objective
To construct a rocket out of a balloon and use it to carry a payload vertically.

- Target concept: Altitude
- Preparation time: 10 min
- Duration of activity: 60 min
- Student group size: Teams of three or four

Materials and Tools
- Large long balloons (several per group)
- Fishing line
- Straws
- Small paper cups
- Paper clips
- Tape
- Clothespins
- Scales

Background Information
The mass of a rocket can make the difference between a successful flight and a rocket that just sits on the launch pad. As a basic principle of rocket flight, a rocket will leave the ground when the engine produces a thrust that is greater than the total mass of the vehicle.

Large rockets, able to carry a spacecraft into space, have serious weight problems. To reach space and proper orbital velocities, a great deal of propellant is needed; therefore, the tanks, engines and associated hardware become larger. Up to a point, bigger rockets fly farther than smaller rockets; however, when they become too large, their structures weigh them down too much.

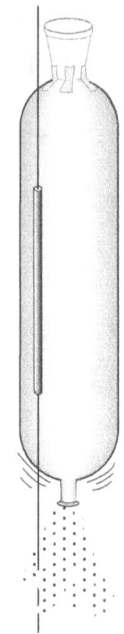

A solution to the problem of giant rockets weighing too much can be credited to the 16th-century fireworks maker John Schmidlap. Schmidlap attached small rockets to the top of big ones. When the large rockets exhausted their fuel supply, the rocket casing dropped behind, and the remaining rocket fired. Much higher altitudes can be achieved this way.

This technique of building a rocket is called staging. Thanks to staging, not only can we reach outer space in the Space Shuttle, but we can also reach the Moon and other planets using various spacecraft.

Procedure
1. Attach a fishing line to the ceiling or as high on the wall as possible. Try attaching a paper clip to a fishing line and hooking it on to the light or ceiling tile braces. Make one vertical drop with the fishing line to the floor or tabletop per group. Thread the fishing line through a straw. Then attach the bottom end of the line to the tabletop or floor.

Note: The fishing line should be taut for the rocket to travel successfully up the line.

2. Blow up the balloon, and hold it shut with a clothespin. You will remove the clip before launch.
3. Use the paper cup as a payload bay to carry the weights. Attach the cup to the balloon using tape. Encourage students to think of creative locations to attach the cup to the balloon. See figure on page 74.
4. Attach the straw to the side of your rocket using the tape. Be sure the straw runs lengthwise along the balloon. This will be your guide and attachment to your fishing line.
5. Launch is now possible simply by removing the clothespin and untwisting before release.
   Note: The line may be marked off in American Standard or metric units with a marker to aid students in determining the height traveled.
6. After trying their rocket, have students predict how much weight they can lift to the ceiling. Allow students to change their design in any way that might increase the rocket's lifting ability between each try (e.g., adding additional balloons, changing locations of the payload bay, or replacing the initial balloon as it loses some of its elasticity, enabling it to maintain the same thrust).

Discussion
1. Compare what you have learned about balloons and rockets.
2. Why is the balloon forced along the string?

Assessment:
Compare results from student launches. Have students discuss design elements that made their launch successful and ideas that they think could be used to create an even more successful heavy lift launcher.

Extensions
- Can you eliminate the paper cup from the rocket and have it still carry paper clips?
- If each balloon costs 1 million dollars and you need to lift 100 paper clips, how much money would you need to spend? Can you think of a way to cut this cost?
- Without attaching the paper cup as a payload carrier, have students measure the distance the balloon travels along the string in a horizontal, vertical and 45 degree angle using American Standard or metric units. Discuss the differences.

Rocket Transportation

Rocket Team: _____
_____
_____
_____

* Predict how much weight your rocket can lift to the ceiling (two small paperclips = approximately 0.04 oz (1g))

Test	Weight Lifted	Results of Test
1		
2		
3		
4		

Based on your most successful launch:

* What was the maximum amount of weight you could lift to the ceiling? _____

Sketch Your Rocket

Explain how you designed your rocket to lift the maximum amount of weight.

_____
_____
_____
_____
_____
_____
_____

* What other ways could increase the lifting capacity of your rocket?

Parachute Area Versus Drop Time

Objective
To study parachute area and its relationship to drop time.
- Target concept: Acceleration
- Preparation time: Less than 10 min
- Duration of activity: 50 min
- Student group size: Teams of three

Materials and tools
- Plastic trash bags or tissue paper
- String
- Payload (plastic army man, pencil, etc.)
- Graph paper
- Rulers
- Tape
- Scissors
- Stopwatch

Management
This lesson can be performed using plastic trash bags or tissue paper for the parachutes. Each group will need enough of the parachute material to make six different parachutes. After the completion of the parachutes, they should be dropped from an elevated area, such as a chair for the student to stand on, or a second floor. Get permission before dropping objects from second stories. Each group will need at least three payloads to attach to their parachutes. Each payload should be the same weight.

Some of the calculations performed in this lesson may be a little too advanced for younger students. If necessary, provide assistance or perform the calculations as a group.

Background Information
Parachutes come in many different shapes and sizes and are made from many different types of materials. Even though they come in a wide variety, they all serve one purpose, to catch air and slow the wearer down for a soft, smooth landing on the ground.

This lesson will give students the opportunity to test different parachute sizes and shapes. They will measure each parachute's area and record the amount of time it takes to reach the ground. This data will then be graphed.

Guidelines
1. Discuss how the parachute works. The parachute creates drag, slowing down the airplane. This added air resistance keeps the plane from plummeting to the ground.
2. Explain that the students are going to be working together to test different parachute designs. They are also going to determine if there is a connection between the area of the parachute and the time it takes the parachute to reach the ground.
3. Place students into groups of two or three, and hand out the Student Sheets.
4. Go over the instructions on the Student Sheets with the class.
5. Hand out the materials and monitor student progress as they work on the activity.
6. If desired, move to the chosen drop location for students to experiment with their parachutes and to collect their data.

Discussion/Wrap-up
- Go over the answers to the questions found at the end of the Student Sheets.
- Discuss the correlation between parachute area and drop time.

Extensions
- Have students create parachutes in the form of other shapes and test them.

Parachute Area Versus Drop Time

Student Sheet(s)

Name_____

Materials
- Plastic trash bag or tissue paper
- String
- Payload (plastic army man, pencil, etc.)
- Graph paper
- Ruler
- Tape
- Scissors
- Stopwatch
- Science journal

Procedure

Your group will be testing parachute shapes and sizes to determine the best combination to safely bring a payload to the ground. Follow the steps below to complete the experiment.

1. To make your first three parachutes, cut out three differently sized squares from the plastic bag or tissue paper. You may need to fold the squares into triangles and trim them to make them exact squares.
2. Using a pencil, label the parachutes A, B and C.
3. Now, determine the area of each parachute.
   a. Measure the length of the sides of the parachute to be sure they are the same.
   b. Use the equation "area of square = length of side × length of side" to calculate the area of each parachute.
   c. Record the data in your science journal.
4. To complete the parachutes, perform the following:
   a. Cut nine pieces of string (each piece should be about 6-in [15.24-cm] long).
   b. Tape the end of a piece of string to each corner of each square parachute.
   c. If extra payload is available, tie the loose ends of the string to the payload.
   d. If necessary, wait until after square parachutes are tested, and use the same payloads for the triangle parachutes.
5. To make your next three parachutes, you will cut three differently sized equilateral triangles from the plastic bag or tissue paper. You may need to fold the triangles to make sure each side is the same length.
6. Using a pencil, label the parachutes 1, 2 and 3.

7. Now, determine the area of each parachute.
   a. Measure the length of the sides of the parachute to be sure they are the same.
   b. Use the equation $(\text{side})(\text{side})\left(\dfrac{\sqrt{3}}{4}\right)$ to calculate the area of each triangular parachute.
   c. Record the data in your science journal.
8. To complete the parachutes, perform the following:
   a. Cut nine pieces of string. Each piece should be about 6-in (15.24-cm) long.
   b. Tape the end of a piece of string to each corner of each parachute.
   c. If extra payload is available, tie the loose ends of the string to the payload.
   d. If necessary, wait until after square parachutes are tested, and use the same payloads for the triangle parachutes.
9. Your teacher will choose an elevated position for you to do your drop tests. To test your parachutes, you must drop each parachute from the exact same location each time.
10. Have one group member drop the parachute while another group member times the parachute's descent. They must time the drop from the moment the parachute is released to the time the payload hits the floor. The third group member can record the data in their science journal.
11. Repeat this process for each parachute. If time permits, drop each parachute twice.
12. Once you have collected all of your data, create a graph comparing the area of the parachute to the drop time. On your graph, label the x-axis with the area of the parachute and the y-axis with the drop time.

Answer the following questions in your science journal.
1. Is there a definite relationship between the area of the parachute and the drop time?
2. Does more area equal more drop time? Why do you think this is so?
3. Were there any variables in this experiment that could have made your results invalid?
4. Does the shape of the parachute make a difference?
5. Which parachute performed the best? Worst? Why do you think so?

Safe Landing

Imagine you are flying along in your airplane. All of the sudden, the engine stops, and you need to bail out. Luckily, you have a parachute to help you land safely. What about the airplane though? Could it use a parachute to help it get back safely too? NASA is working with a few companies to design a parachute for airplanes in trouble.

The new parachute will allow the endangered airplane and its passengers to land together. This means that in an emergency, passengers would not have to bail out of the aircraft. Pilots and passengers will not have to learn how to use individual parachutes. Instead, they can remain fastened in their seats for a safe landing.

Many different things can cause an airplane to crash land. Engine failure, midair collision, out-of-control spin, pilot error, running out of fuel, and even ice on the wings can lead to disaster. The plane parachute could be used in any of these situations. At the first sign of trouble, the pilot shuts off the engine and pulls a handle. Pulling the handle ignites a rocket that deploys the parachute. This feels

Photograph provided by Ballistic Recovery Systems

like slamming on the brakes in your car. Once the parachute is fully open, the aircraft descends to the ground at the safer rate of 25 ft/s (7.6 m/s).

The rocket that deploys the parachute is a smaller version of the Space Shuttle's solid rocket boosters. It is released from a special opening on top of the plane's fuselage. The parachute is deployed at speeds over 100 m/hr (160 k/hr). It is connected to the airplane with Kevlar® straps. These straps help to keep the plane level. The parachute rescue system can safely deploy at altitudes as low as 300 ft (91 m).

In the early stages, the parachute system was designed for lightweight aircraft such as the ultralight plane. As the technology improved, other small aircraft were also equipped with the parachutes. Today, the system is used on planes that hold up to four passengers.

The next step is to develop the parachute system for planes that carry up to 12 people. One challenge is making the system compact and lightweight. The entire apparatus can weigh no more than 60 lb (27.22 kg). Engineers are testing a new lightweight material for the parachute. It is a reinforced film that has been used for high-performance boat sails.

So far, the plane chute has had over 150 success stories. It has saved many lives and small aircraft. Just last year, a pilot flying in a single-engine plane had to make an emergency landing with the plane chute. He landed safely in a tree grove. The pilot was uninjured, and the plane only suffered minimal damage.

The Parachute Recovery System is a team effort with Ballistic Recovery Systems. They are working together as part of NASA's Small Business Innovation Research program. This program's goal is to make NASA technology available to smaller companies.

Kevlar is a registered trademark of E.I. du Pont de Nemours and Company.

Balloon Staging

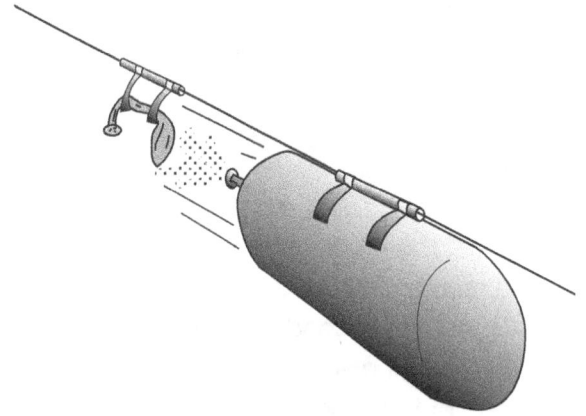

Objective
To simulate a multistage rocket launch by using two inflated balloons that slide along a fishing line by the thrust produced from escaping air.

- Target concept: Velocity
- Preparation time: 10–30 min
- Duration of activity: 55–65 min
- Student group size: Teams of two or three

Materials and Tools
- Two long party balloons
- Nylon monofilament fishing line (any weight)
- Two plastic straws (milkshake size)
- Styrofoam coffee cup
- Masking tape
- Scissors
- Two spring clothespins

Management
The activity described below can be done by students or used as a demonstration. Younger students may have difficulty in coordinating the assembly steps to achieve a successful launch. For safety, consider attaching the fishing line along one wall where there is not much traffic, so students will not walk into the line.

Background Information
Traveling into outer space takes enormous amounts of energy. This activity is a simple demonstration of rocket staging that Johann Schmidlap first proposed in the 16th century. When a lower stage has exhausted its load of propellants, the entire stage drops away, making the upper stages more efficient in reaching higher altitudes. In the typical rocket, the stages are mounted one on top of the other. The lowest stage is the largest and heaviest.

In the Space Shuttle, the stages attach side by side. The Solid Rocket Booster's attach to the side of the external tank. Also attached to the external tank is the Shuttle orbiter. When exhausted, the SRBs jettison. Later, the orbiter discards the external tank as well. Thanks to staging, not only can we reach outer space in the Space Shuttle, but we can also reach the Moon and other planets using various spacecraft.

Procedure
1. Thread the fishing line through the two straws. Stretch the fishing line snugly across a room, and secure its ends. Make sure the line is just high enough for people to pass safely underneath.

83

2. Cut the coffee cup in half so that the lip of the cup forms a continuous ring.
3. Stretch the balloons by preinflating them.
    a. Inflate the first balloon about three-fourths full of air and squeeze its nozzle tight.
    b. Pull the nozzle through the ring.
    c. Twist the nozzle and hold it shut with a spring clothespin.
    d. Inflate the second balloon.
    e. While doing so, make sure the front end of the second balloon extends through the ring a short distance. As the second balloon inflates, it will press against the nozzle of the first balloon and take over the clip's job of holding it shut. It may take a bit of practice to achieve this.
    f. Clip the nozzle of the second balloon shut also.
4. Take the balloons to one end of the fishing line and tape each balloon to a straw with masking tape. The balloons should point parallel to the fishing line.
5. Remove the clip from the first balloon and untwist the nozzle. Remove the nozzle from the second balloon as well, but continue holding it shut with your fingers.
6. If you wish, do a rocket countdown as you release the balloon you are holding. The escaping gas will propel both balloons along the fishing line. When the first balloon released runs out of air, it will release the other balloon to continue the trip.
7. Distribute design sheets and ask students to design and describe their own multistage rocket.

Assessment
1. Collect and display student designs for multistage rockets.
2. Ask each student to explain his/her rocket to the class.

Extensions
1. Encourage the students to try other launch arrangements, such as side-by-side balloons and three stages.
2. Can students fly a two-stage balloon without the fishing line as a guide? How might the balloons be modified to make this possible?

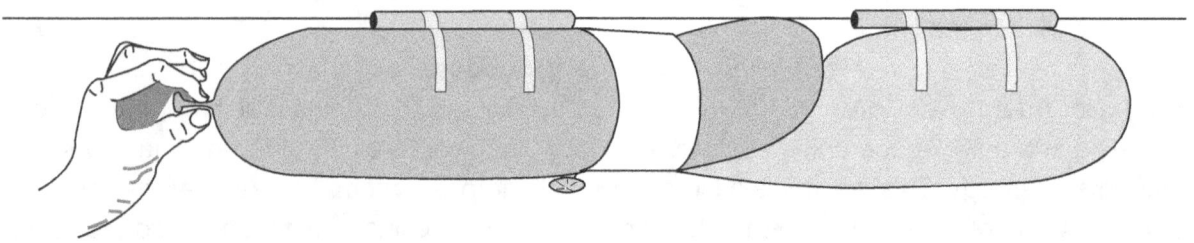

Water Bottle Rocket Assembly

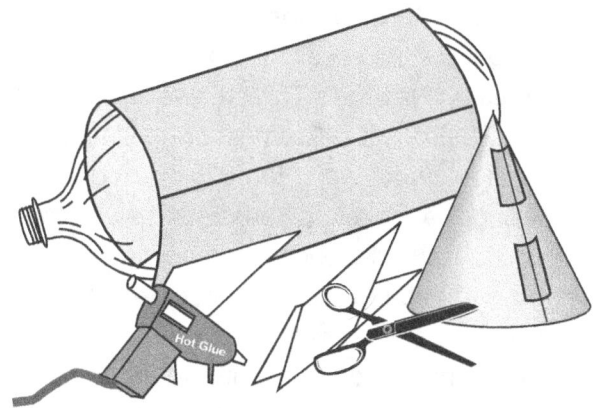

Objective
Working in teams, learners will construct a simple bottle rocket from 2-liter soft drink bottles and other materials.

- Target concept: Altitude, velocity and acceleration
- Preparation time: Less than 10 min
- Duration of activity: 50 min
- Student group size: Teams of three students

Materials and Tools
- 2-liter plastic soft drink bottles
- Low-temperature glue guns
- Poster board
- Tape (masking and/or duct)
- Modeling clay
- Scissors
- Safety Glasses
- Decals
- Stickers
- Marker pens
- Bottle rocket launcher (see page 88)

Management
Having the learners work in teams will reduce the amount of materials required. Begin saving 2-liter bottles several weeks in advance to have a sufficient supply for your students. You will need to have at least one bottle rocket launcher. Construct the launcher described on page 146 or obtain one from a science or technology education supply catalog or on-line.

Collect a variety of decorative materials before beginning this activity so students can customize their rockets. Rockets can be constructed with the materials and attached by tape or low-temperature electric glue guns that are available from craft stores. High-temperature glue guns will melt the plastic bottles. Provide tape or glue guns for each table or set up glue stations in various parts of the room. When the rockets are complete, test fly them.

When launching rockets, it is important for the other students to stand back. Countdowns help everybody to know when the rocket will lift off. In a group discussion, have your students create launch safety rules that everybody must follow. Include how far back observers should stand, how many people should prepare the rocket for launch, who should retrieve the rocket, etc.

Refer to the Altitude Tracker activity starting on page 109 for information on determining how high the rockets fly. While one group of students launches its rocket, have another group track the rocket and determine its altitude.

Background Information

Bottle rockets are excellent devices for investigating Newton's Three Laws of Motion. The rocket will remain on the launch pad until an unbalanced force is exerted propelling the rocket upward (First Law). The amount of force depends upon how much air you pumped inside the rocket (Second Law). You can increase the force further by adding a small amount of water to the rocket. This increases the mass the rocket expels by the air pressure. Finally, the action force of the air (and water) as it rushes out the nozzle creates an equal and opposite reaction force propelling the rocket upward (Third Law).

The fourth instruction on page 87 asks the students to press modeling clay into the nose cone of the rocket. Placing 1.75–3.53 oz (50–100 g) of clay into the cone helps to stabilize the rocket by moving the center of mass farther from the center of pressure. For a complete explanation of how this works, see pages 11–14. An example of the nose cone can be obtained from page 39.

Procedures

Refer to the page 87 for procedures on building a water bottle rocket.

Assessment

- Evaluate each bottle rocket on its quality of construction.
- Observe how well fins align and attach to the bottle.
- Observe how straight the nose cone is at the top of the rocket.
- If you choose to measure how high the rockets fly, compare the altitude the rockets reach with their design and quality of construction.

Extensions

- Challenge rocket teams to invent a way to attach a parachute to the rocket that will deploy on the rocket's way back down.
- Parachutes for bottle rockets can be made from a plastic bag and string. The nose cone is merely placed over the rocket and parachute for launch. The cone needs to fit properly for launch or it will slip off. The modeling clay in the cone will cause the cone to fall off, deploying the parachute, after the rocket tilts over at the top of its flight.
- Conduct flight experiments by varying the amount of air pressure and water to the rocket before launch. Have the students develop experimental test procedures and control for variables.
- Conduct spectacular nighttime launches of bottle rockets. Make the rockets visible in flight by taping a small-size chemical light stick near the nose cone of each rocket. (Light sticks are available at toy and camping stores and can be used for many flights. This is an especially good activity for summer space camp programs.)

Building A Bottle Rocket

1. Wrap and glue or tape or a tube of posterboard aroun the bottle.

2. Cut out several fins of any shape and glue them to the tube.

3. Form a nose cone and hold it together with tape or glue.

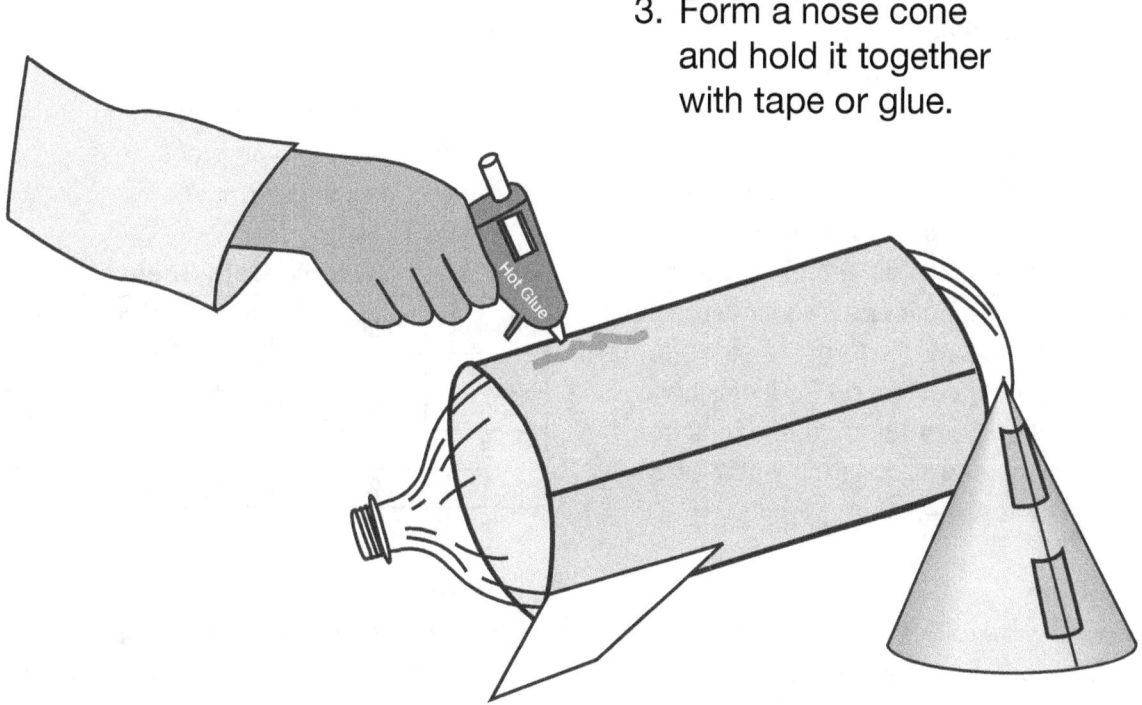

4. Press a ball or modeling clay into the top of the nose cone.

5. Glue or tape nose cone to upper end of bottle.

6. Decorate your rocket.

Launching Water Bottle Rockets

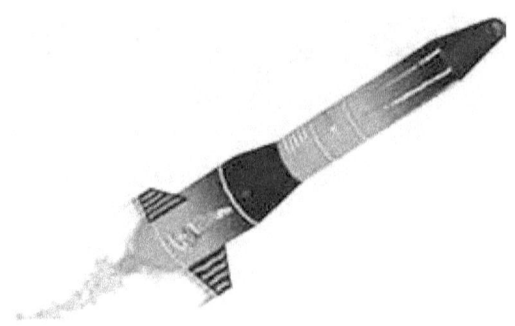

Objective
To show the connection of propulsion and force, pushing the rocket by accelerating water out of a bottle. Expand on the straw rocket exercise by building a water rocket that will fly powered by pressurized water.

- Target word: Acceleration
- Preparation time: 1 hour
- Activity time: 1 hour
- Students involved: 3–12 per adult

Preparation
Refer to the Safety Code in Appendix B. Build the water bottle rockets in three or more teams of three students each. Materials and assembly instructions are given in a previous section on page 85. Perform the string-swing test given in the section on Rocket Stability Determination page 138 on these rockets. You will also need to build the water bottle launcher (see Appendix A).

Procedure
Students will explore what relative amounts of air and water propel the rocket better, or make any design adjustments to the fins or nose. Observing the flight of various lengths and weights of rockets will add to the inquiry experience.

1. Begin by having some bottles filled one-fourth, one-third and one-half full of water.
2. Set them on the launcher, and pump these rockets each with the same amount of air (same number of pumps). Then, launch them to see which ones go the highest or furthest.
3. Note the results on a data sheet.
4. Then, select a set water amount and change the number of pumps to increase or decrease the air pressure for the next launch round.
5. Note the results on a data sheet.
6. Determine the best combination of the water fill level and air pressure, and fly all rockets with that formula.

Sample Data Sheet

Rocket	Flight	Water Fill	Air Pump Strokes	Height
1	1	1/2	4	low
2	1	1/3	4	low
3	1	1/3	6	high
1	2	1/4	8	highest

Questions
1. How is the air and water pressure operation better than just the air pressure with the straw rocket?
2. What if you just put air in the bottle? Just air pressure drove the straw rocket in a pulse. The bottle has the pressure expelling water. Ask them to explain the difference. The action of expelling the water mass causes the reaction of the bottle motion (Newton's Third Law).
3. How would a 2-liter bottle fly differently from a half-liter bottle? Hypothesize what would happen and then build and launch a rocket made from a small bottle.

4. How would you relate pressure, volume and bottle mouth size against rate of water expulsion for effective propulsion?
5. How do you relate the operation of this water-propelled rocket to a liquid fuel combustion propelled rocket?

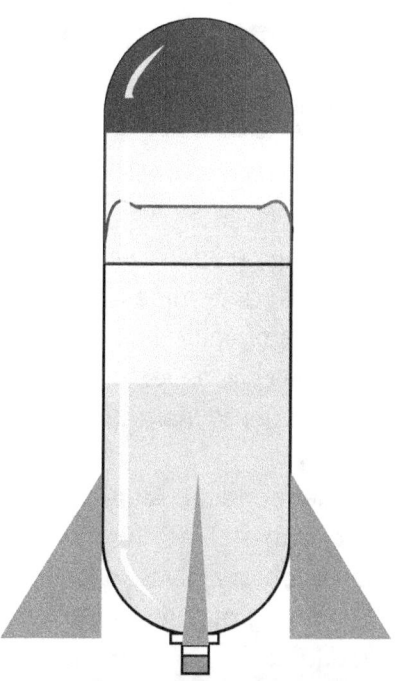

The Nose Cone Experts

Objective
Students will experiment with different nose cone shapes to determine the advantages and disadvantages of each type. Conic, parabolic and flat shapes will be tested to determine which is most aerodynamic.

- Target concept: Altitude
- Preparation time: 10 min
- Duration of activity: 45–60 min
- Student group size: One to two students

Materials and tools:
- Nose Cone Distance Traveled Table
- Group Questions and Procedures
- Paper towel tube
- Nose cone patterns worksheet
- Yard stick or meter stick
- Several 2-liter plastic soft drink bottles
- Modeling clay
- Card stock
- Leaf blower or vacuum set to blow
- Books to make a path
- Long hall or open area
- Tape

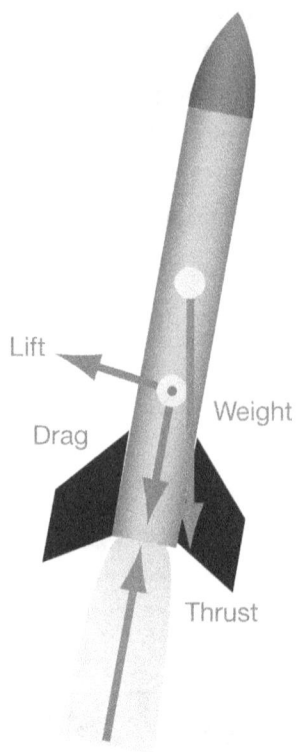

Forces on a Rocket

Background
Aerodynamics is the branch of science that deals with the motion of air and the forces on bodies moving through the air. There are four forces that act on a rocket. They are lift, drag, weight and thrust.

Drag is a force that opposes the upward movement of the rocket. It is generated by every part of the rocket. Drag is a sort of aerodynamic friction between the surface of the rocket and the air. Factors that affect drag include the size and shape of the rocket, the velocity and the inclination of flow, and the mass, viscosity and compressibility of the air.

Procedure
1. Students in this expert group will complete the What a Drag prelab sheet to allow them to build on their past experiences with aerodynamics.
2. Students construct nose cones by cutting out three different nose cone shapes from card stock. Two of the patterns are given on the following pages. The third nose cone should be a flat against the paper towel tube. They will then attach the nose cones onto paper towel tubes. This time modeling clay can be used inside the nose cone to provide mass.
3. Use a commercial leaf blower or a vacuum cleaner with its airflow reversed to blow to force the rocket backwards. This should

be done on a narrow track to keep the rocket in line with the wind (books may be lined up to make this track).
4. Students should measure the distance the rocket traveled backwards. Record the results and complete the nose cone expert report on the What a Drag sheets.

Group Questions and Procedure

In your expert groups, complete the following:
1. What is the first thing you think of when you hear the word aerodynamic? Where have you heard the term before?
2. Using the resources on the Internet or in your library, find information on aerodynamics and the importance of the use of wind tunnels. Give several examples.
3. What is drag as it relates to aerodynamics?
   a. What are some things that can be done to an object to decrease its drag?
   b. What are the parts of a rocket that may result in drag?
4. Using the patterns in the nosecone pattern, cut out three different nose cone shapes from card stock. Assemble the nose cones onto paper towel tubes. The tubes will be tested with the leaf blower as shown below.
5. List the variables that need to be controlled in this activity.
6. Use a commercial leaf blower or vacuum set to blow air to force the rocket backwards. This should be done between two rows of books to keep the rocket in line with the wind.
   a. Place the nose cone design in front of the blower, as shown below.
   b. While holding the blower, turn the blower on until the nose cone design stops moving.
7. Measure the distance the rocket traveled backwards.
8. Record results in the data table below.
9. Set up a data table in your journal similar to the table on page 92 to record your results.

Nosecone Distance Traveled Table

Shape of Nosecone	Distance Traveled			
	Trial One	Trial Two	Trial Three	Average

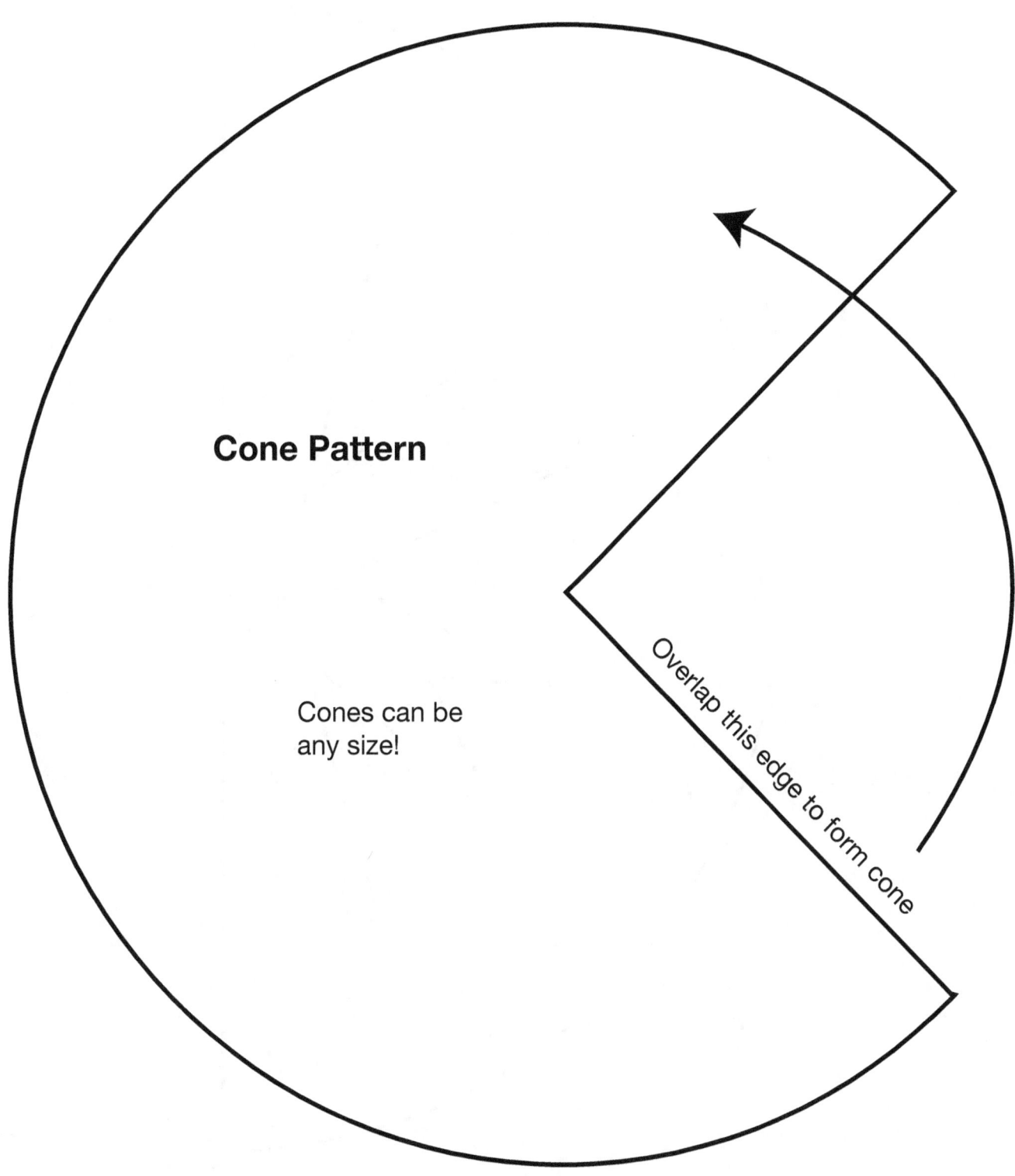

93

Cut along the lines. Tape the sides of the triangles together starting with the smaller triangles.

Racing Against Friction

Objective
To understand how friction affects the speed of a vehicle.

- Target concept: Velocity
- Preparation time: 20 minutes
- Duration of activity: 40–45 minutes
- Student group size: Teams of two to four students

Materials and Tools
- Large sheets of corrugated cardboard
- Masking tape
- Felt fabric
- Wax paper
- Sandpaper
- Construction paper
- Various textbooks
- Small toy cars
- Stopwatches
- Student Sheets
- Scissors

Management
Before the activity begins, cut out strips of felt fabric, wax paper and sandpaper slightly wider than the width of a toy car and approximately 1 ft (30.48 cm) long. Ensure each group has a piece of cardboard approximately 1.5 × 2 ft (45.72 × 60.96 cm) to make their ramp surface. Read Two-Ton Hockey Pucks on page 98 to the students.

Background Information
Working in space can be tricky. With no gravity or friction to keep things in place, relatively simple tasks can become complicated ordeals. To prepare for the rigors of

working in space, astronauts train in many different facilities on Earth. One of these facilities, the Precision Air Bearing Facility at Johnson Space Center in Houston, Texas, is used to simulate the reduced friction found in space.

This lesson will introduce students to the concept of friction being a slowing force.

Procedure
1. Write the word FRICTION on the board. Have students share any information they may know about friction.
2. Explain that the class is going to investigate friction and the effects it has on a moving vehicle.
3. Place students into groups and hand out the Student Sheets.
4. Go over the instructions on the Student Sheets and answer any questions the students may have.
5. Allow time for the students to complete the activity.

Discussion/Wrap-up
Have students share their results, and discuss why the results turned out as they did.

Extensions
Study Isaac Newton's Laws of Motion.

Racing Against Friction

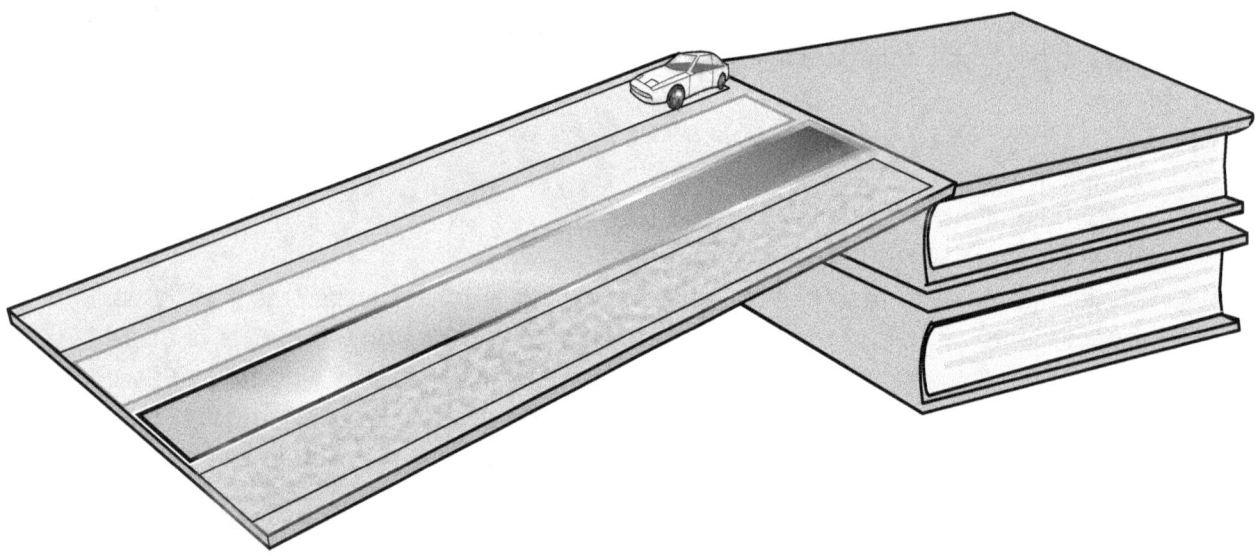

Procedure

1. Use the scissors to trim the different strips of material to the same length.
2. Place the strips of material on the piece of cardboard. One end of each strip should be lined up against the edge of one side of the cardboard. (See the above diagram.) Tape the strips in place using the masking tape.
3. Stack the textbooks on top of one another. Place one end of the cardboard on top of the books to form a ramp. The ends of the strips of material should be toward the table. Tape the cardboard in place.
4. Predict which material will allow the car to move down the ramp the quickest. Write your prediction on the Data Sheet, and explain your prediction.
5. Place the toy car at the top edge of the first strip of material. Let the car roll down the ramp to the table. Use the stopwatch to time the amount of time it takes the car to travel from the top of the material strip to the table. Record the time on the Data Sheet.
6. Repeat this process with the first strip of material until you have completed three trials. Record all data.
7. Repeat steps 5 and 6 with the other three strips of material. Record all results.
8. Answer the questions on the following Data Sheet.

Racing Against Friction Data Sheet

Name _____

Fastest material prediction _____

Explain your choice _____

_____

_____

_____

_____

Data Table

	Trial One Time (s)	Trial Two Time (s)	Trial Three Time (s)	Average Time (s)
Construction paper				
Felt fabric				
Sand paper				
Wax paper				

Questions
1. Which material was the fastest track for the toy car? Was your prediction correct?
2. Why did the toy car travel at different rates on the different materials?
3. Why is it important for the strips of materials to be the same length?

Two-Ton Hockey Pucks

Superman is not the only one who can move a 2-ton (1,814.37-kg) automobile with his pinkie finger. Astronauts can do it too. There are not any cars in space, but the crew can move objects just as heavy with just as much ease. In microgravity, there is not any friction to provide resistance, so large objects take off with just a nudge. The only problem is that, while it is easy to glide large pieces of equipment, they can be hard to maneuver or stop. Nobody wants a 2-ton (1,814.37-kg) object drifting out of control. In order to keep things under control in space, astronauts practice working in microgravity on Earth.

That is where it gets tricky. There is not one type of microgravity simulator that perfectly duplicates the conditions of space. That is why NASA has developed a range of simulators; each one recreates a specific aspect of microgravity. These simulators include the KC-135 airplane that allows astronauts to tumble through the air in a free fall. The Neutral Buoyancy Laboratory takes astronauts to an underwater lab almost like space. Virtual reality simulators reproduce the visual aspects of space, and the microgravity drop towers briefly put objects and experiments in a near-microgravity setting. But when it comes to simulating how to move those large objects, performing functions without friction and responding to the lack of gravity's counter forces, astronauts head to the Precision Air Bearing Facility at Johnson Space Center in Houston.

There, they take advantage of the air-bearing floor. Imagine a giant air hockey table about 20 ft (6 m) wide and 30 ft (9.1 m) long, says Tom Smith, mock-up manager for the Precision Air Bearing Facility. Once on the special floor, astronauts can push huge objects around as easily as an air hockey puck floats across a game table, which is how it feels in space. Smith says that if you were playing air hockey in a game room, the hockey puck would float on a thin cushion of air blown from the surface of the table. However, on the air-bearing floor, the air is essentially fed through the hockey puck to create a layer of compressed air hovering just fractions of an inch (millimeters) off the floor. Needless to say, in order for 2-ton (1,814.37-kg) hockey pucks to maintain a position so close to the surface, that surface must be very smooth and level.

The floor is made up of steel plates, lined up side by side. More steel plates are combined to form pads; these pads are attached to the bottoms of the large objects astronauts practice moving. From there, compressed air

is fed into a tube that runs along the sides and bottom of the plates. When running, the air-fed plates hover just above the floor's surface, and all it takes is a nudge for mammoth objects to effortlessly glide across the surface.

What kind of chores do astronauts practice on the air-bearing floor? They primarily handle large objects—as large as a small car—and learn to keep the objects moving in their intended direction. They also practice using tools the way they would be used in space. Without gravity to keep the astronaut in place, a simple twist of a screwdriver can send him or her zooming. Astronauts can compensate for the torque created by drills, wrenches and pliers by anchoring themselves or bracing their bodies.

"Engineers also use the air-bearing floor as they are developing products and parts that will travel into space," Smith says. A simple device like a door hinge can be tested to see how it will function without the friction found on Earth. By operating the hinge on the giant air hockey table, it is easy to see if changes need to be made because of microgravity.

"Some people call this a zero-gravity room, but that is not accurate," says Smith. "There is plenty of gravity in here. This floor simulates the effects of reduced gravity on one plane only, not throughout the entire room. It is pretty amazing, though, to see how easy it is to move something weighing 2 tons (1,814.37-kg) with just your pinkie."

The Parachuting Egg

Objective
To drop an egg from a height without it breaking.

- Target concept: Acceleration
- Preparation time: 10 min
- Duration of activity: 50–60 min
- Student group size: Teams of two to five

Background Information
After the Space Shuttle is launched, the Solid Rocket Boosters (SRBs) are jettisoned at about 2 min into the flight. SRB separation occurs at an altitude of about 30 miles (48.3 km). The separated boosters then coast up to an altitude of 47 mi (75.6 km) and free-fall into an impact zone in the ocean about 158 mi (254.3 km) downrange. They are retrieved from the Atlantic Ocean by special recovery vessels and returned for refurbishment and eventual reuse on future Shuttle flights.

When a free-falling booster reaches an altitude of about 3 mi (4.8 km), its nose cap is jettisoned, and the SRB pilot parachute pops open. The pilot parachute then pulls out the 54 ft (16.5 m) diameter drogue parachute. The drogue parachute stabilizes and slows down the descent to the ocean. At an altitude of 6,420 ft (1,902 m), the frustum, a truncated cone at the top of the SRB where it joins the nose cap, is separated from the forward skirt, causing the three main parachutes to pop out. These parachutes are 115 ft (35 m) in diameter and have a dry weight of about 1,500 lb (680.4 kg) each. When wet with seawater, they weigh about 3,000 lb (1,360.8 kg).

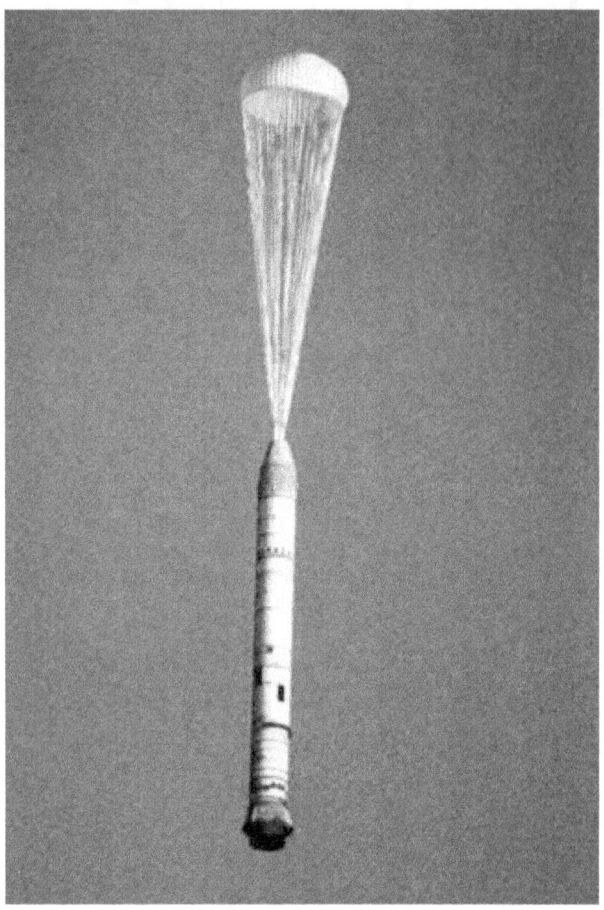

At 6 minutes and 44 seconds after liftoff, the spent SRBs, weighing about 165,000 lb (74.8 metric tons), have slowed their descent speed to about 62 mph (99.8 kph), and splashdown takes place in the predetermined area. The parachutes remain attached to the boosters until recovery personnel detach them. Under ideal weather and sea conditions, the retrieval operation takes about 5.5 hours. The recovery ships with the retrieved SRBs in tow, sail to Port Canaveral, travel north up the Banana River, and dock near Hangar AF at the Cape Canaveral Air Force Station, their mission completed.

Once the parachutes have been recovered, they are taken to the Parachute Refurbishment Facility (PRF) in the Kennedy Space Center. The PRF was originally built to process the

parachutes used in the Gemini manned space program and was modified for the Shuttle program. Parachute systems recovered with the SRB casings are delivered to the refurbishment facility on reels provided on the retrieval vessels. The parachutes for the SRBs are washed, dried, refurbished, assembled and stored in this facility. New parachutes and hardware from manufacturers also are delivered to the PRF.

Parachutes arriving from the retrieval operations are untangled, hung systematically on an overhead monorail and transported to in-line washers and driers that wash and dry them automatically. They are then sent to the refurbishing area, which is equipped with handling equipment for producing and installing risers and associated equipment. After final inspection and acceptance, the chutes are folded and placed in canisters for reuse.

Materials
- Egg provided by your group
- Parachuting materials brought in by your group
- Stopwatch
- Tape measure

Procedure
1. Your group will design a parachute that can be attached to an egg. The egg will be dropped from a known height to the ground or floor. The parachute is to slow the egg's descent so that, after landing, it remains uncracked.
2. You will spend time in class today discussing how you will design and build your parachute. It can be made out of anything you choose. Due to the quickness that this has to be built, try using household items.
3. The parachute's length, including any strings or drag devices, can only be 3.28 ft (1 m).
4. The parachute must allow the egg to come down. This is not a flying contest, but a safe, dropping contest.

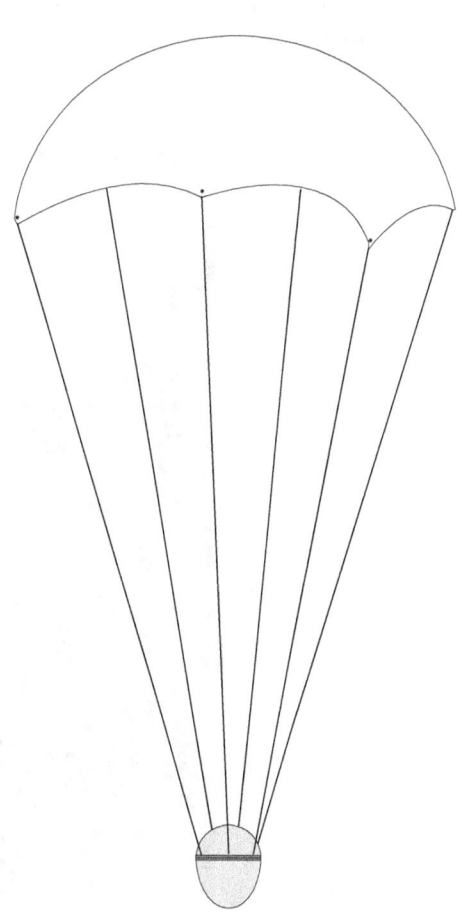

5. There are no set criteria for your parachute, unless dictated by your teacher due to physical or time constraints.
6. Bring in a hard-boiled or raw egg, based on what your teacher says.
7. Be sure it can be seen from the outside when attached to your parachute. There should be no foreign material wrapped around the egg to protect it from the fall (paint, tape, glue, finger nail polish, paper, bubble wrap, styrofoam, straws, springs, popsicle sticks, etc.). The design of this experiment is to make a working parachute, not an egg protection suit.
8. You can build a harness for it to sit in, so long as the egg hits the ground first and not the harness. The harness should not provide any extra stability for the egg, other than to hold it to the parachute.
9. During the drop, you will need to measure the time from the release to the impact.
10. If it is not already provided, you will need to measure the vertical distance the egg fell.
11. Using the speed equation, determine the speed of your egg when it hit the ground. Answer the following questions:
    a. What is your basic design and how does it work? Include a small sketch of your parachute.
    b. In designing your parachute, what were some of the factors that you had to account for to ensure a safe landing?
    c. How far did your egg fall?
    d. How long did it take to fall?
    e. What was the speed of your egg at impact?
    f. Did the egg break? What does that tell you about your speed at impact?
    g. How could you improve the safety of your egg? Explain.

Egg Drop Lander

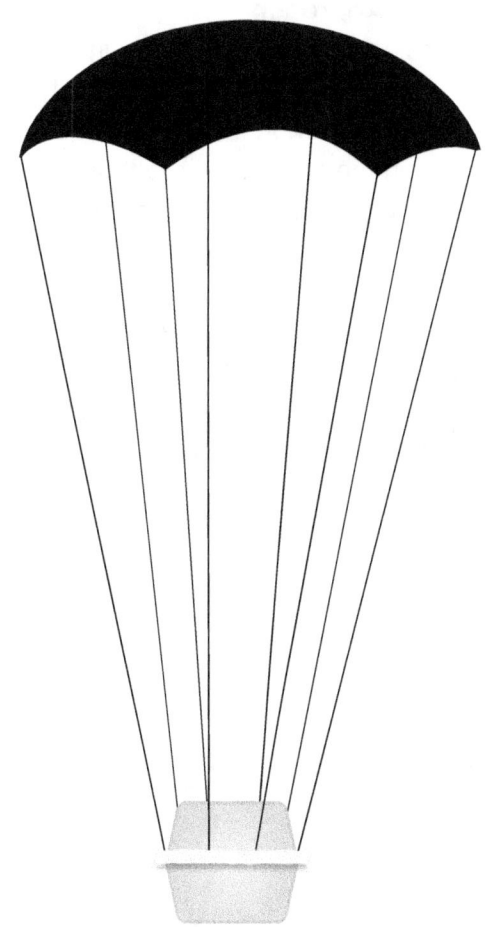

Objective
Students will create a package to contain and successfully land a raw egg, unbroken from a fall to the ground. They will learn how velocity and acceleration from falling objects relate to a force on landing.

- Target concept: Acceleration
- Preparation time: 1 hour
- Activity time: 1 hour
- Student group size: Teams of three (3 to 12 per adult)

Materials
- Raw egg
- Parachute material (plastic trash or shopping bags)
- Packing material (gelatin, popcorn, foam, bubble wrap, etc.)
- Masking tape
- Yardstick or meter stick
- Stopwatch

Procedure
1. Each team of three students will build its own lander capsule. You may wish to build more than one for experimentation. Select someone to be a timekeeper, distance measurer and data recorder.
2. Choose the parachute and packaging material you will use around the egg. Design and build your lander. Attach the parachute.
3. The landing site will be a 1×1 ft target.
4. From the top of a ladder over the target, drop your lander. A balcony is a good place to use too.
5. Record the distance and time it takes for the egg lander to reach the ground.
6. Examine and record the lander. A drop is successful if the egg does not crack.

Data and Results
1. List the packaging material used. Which material and packing technique worked the best?
2. Draw your design.
3. Time of the fall_____s
4. Distance of the fall_____ft (m)
5. At what speed did the box hit the ground: ft/s (m/s)? _____
   (speed = distance/time or ft/s (m/s)

Additional Approach

From what you learned in packaging and protecting the egg in this lander drop test, design a capsule from a model rocket nose cone that can contain the egg. Test drop that capsule to prove the egg in it can land safely. There are also commercial rocket kits that can carry eggs. Get one of those as a design comparison and fly it, then have students build their own version of an egg-carrying rocket with their capsule. Launch the egg in the rocket and see how well the parachute brings it down.

At the Drop of a Ball

Objective
To recreate Galileo's experiment and determine the acceleration of gravity.

- Target concept: Acceleration
- Preparation time: 10–30 min
- Duration of activity: 45 min
- Student group size: Individually or in pairs

Materials
- Golf ball
- Softball
- Basketball
- Stopwatch
- Tape measure
- Calculator

Background Information
It may be the most famous physics experiment ever done. Four hundred years ago, Galileo Galilei started dropping things off the Leaning Tower of Pisa and timing their falls. In those days, it was widely thought that heavy objects fell faster than light ones, but he found that everything hit the ground at the same time. For cannon balls, musket balls, gold, silver and wood, gravity accelerated each item downward at the same rate, regardless of mass or composition.

Today's experiment will have your group recreating Galileo's experiment. The size of the golf ball will be larger, but similar in weight to a musket ball (in other words, a bullet in Galileo's time). The softball will be lighter in weight, but of equivalent size to a cannon ball. The place of the experiment is

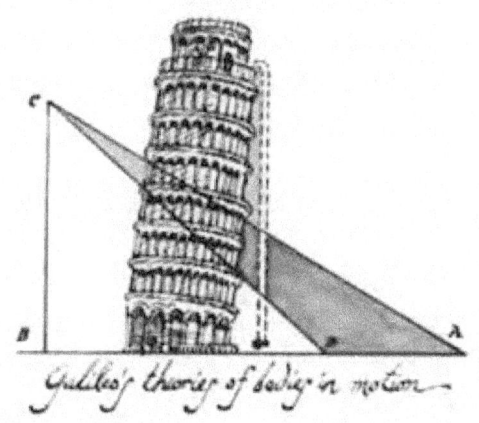

not important, but the Leaning Tower of Pisa did provide a clear view of objects dropped straight down. So, what do you think? Will a softball and a golf ball dropped at the same time, fall to the ground at the same rate and hit the ground at the same time?

For advanced students: Here is a quick review of the equations of motion, which you will need today. The first one is $v_f = v_i + at$, where v_f is the final velocity (ft/s or m/s), v_i is the initial velocity (ft/s or m/s), a is the acceleration in feet (or meters) per second squared (ft/s^2 or m/s^2), and t is the time in seconds (s). The second one is $d = v_i t + 0.5at^2$, where d is the distance traveled in feet (or meters). The third one is $v_f^2 = v_i^2 + 2ad$. For falling objects, we usually substitute $-g$ for a, where $g = 32.1$ ft/s^2 (9.8 m/s^2), and y for d, since the distance traveled is linear in the
vertical direction only.

Procedure
1. Acquire a golf ball, a softball and a stopwatch.
2. Travel to your "Leaning Tower of Pisa" (window or balcony) for your experiment.

3. Measure the height above the ground (drop zone) that you will be dropping the golf ball and softball. Convert this to feet (meters), and record the number in the chart.
4. Have one group member climb to the top of the Leaning Tower to drop the bullet and cannon ball.
5. Have one group member time the event in seconds, and have another group member record the results.
6. Repeat the experiment several times and average the results in the chart.
7. Using the equations of motion, determine the final velocity (the velocity just before it hits the ground) and acceleration of the softball and golf ball.
8. Acquire a basketball to repeat the experiment with the softball and golf ball. Record your results, and use the equations of motion to determine the final velocity and acceleration.
9. Answer the following questions:

a. Did the golf ball and softball hit the ground at the same time?
b. What are some of the things that this experiment proved?
c. Did the basketball change the outcome?
d. What were the sources of error for this experiment?
e. How could you improve the accuracy of this experiment?
f. How does this experiment compare to the way Galileo did it 400 years ago?
g. If you were to compare a piece of paper and a golf ball, which would fall faster? Would they hit the ground at the same time? Why or why not? (If time allows, test it.) (Hint: What is terminal velocity?)
h. Average the acceleration for the five trials using the softball and golf ball. Record that number here:_____. What value should this be? Determine the percent difference between your value and the expected value.

Drop of a Ball recording chart

Experiment	Distance of fall (ft or m)	Time (s)	Final Velocity (ft/s or m/s)	Acceleration (ft/s^2 or m/s^2)
Softball and golf ball				
Softball and basketball				
Golf ball and basketball				

Free Fall Rocket Ball Drop

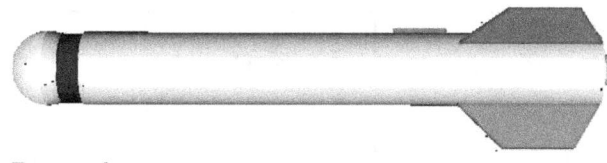

Objective
To calculate the altitude of a rocket by timing the fall of a ball as it drops. This exercise will teach the relation of free fall time and altitude from acceleration.

- Target concept: Acceleration, altitude
- Preparation time: 30 min
- Duration of activity: 60 min per 10 students
- Student group size: Team of 1–3 students (3–12 students per adult)

Materials
- Golf balls, possibly painted bright colors
- A model rocket with a BT60 type body diameter (1.6 in [4.1 cm]) or made from a paper towel tube to contain a golf ball as a nose cone
- Motors to launch the rocket to between 100 and 400 ft (30.48 and 121.92 m) altitude, B or C size
- Launch pad and firing system
- Stopwatches
- Masking tape

Procedure
1. Each student can have their own rocket or work in teams of three. A launch pad operator and timekeeper are also needed.
2. Modify the rocket by removing the nose cone from the shock cord and connect the parachute directly to the shock cord.
3. Wrap two layers of 0.75-in (1.91 cm) masking tape around the ball to form a shoulder so that it can fit snugly into the top of the tube. It should have the same fit tightness as the nose cone did, snug enough not to fall out if turned upside down, but able to be popped off by the ejection charge.
4. Prepare the rocket for flight as usual, but place the golf ball on top of the parachute in place of the nose cone.
5. Launch the rocket.
6. At apogee, the golf ball will eject from the rocket and free fall to the ground. The rocket will descend normally on the parachute. Have a student use a stopwatch to time the ball's fall from the ejection point where the ball separates from the rocket in the air to the moment of ground impact (2–5 seconds).
7. Record the time measured on a data sheet (see page 108).
8. Use the time value to solve for the free fall distance ($y = 0.5gt^2$) (where $g = 32$ ft/s^2 or 9.8 m/s^2), which is the presumed peak altitude of the rocket.
9. Repeat the experiment several times, or have students launch balls in identical rockets and motors and compare the measurements and calculations. Using two stopwatches and taking the average time will improve accuracy and also helps ensure that you get a time reading if one of the watches fails during the event.

Sample data sheet

Flight	Ball drop time (t) (s)	Altitude (y) = 0.5gt² (ft or m)
Model 3 on B6-4	3.3	174

Questions
1. What aspects of the golf ball are being ignored in the descent measurement?
2. If a billiard ball or baseball were dropped instead of the golf ball, how much faster would they fall?
3. What would be a problem with using a ping-pong ball for this experiment?
4. If the ball separates from the rocket with a lot of sideways motion, how does that affect the descent time of the fall?

Additional approach
As you launch, have other students measure the altitude with the altimeter from that activity lesson on page 109. Compare these altitude values. If you have a rocket altitude simulation software program, how does the prediction compare to the calculated value for the rocket?

You can also launch the rockets with different size motors to measure the different altitudes to compare the motor performance effects. Another student can time the ascent of the rocket (lift-off to ejection) and compare it to the ball's free fall time.

Altitude Tracking

Objective
To estimate the altitude a rocket achieves during flight.

- Target concept: Altitude
- Preparation time: 20 min
- Duration of activity: 60 min
- Student group size: Teams of two

Materials and Tools
- Altitude tracker pattern
- Altitude calculator pattern
- Thread or lightweight string
- Small washer
- Brass paper fastener
- Scissors
- Razor-blade knife and cutting surface
- Stapler
- Yard stick or meter stick
- Rocket and launcher

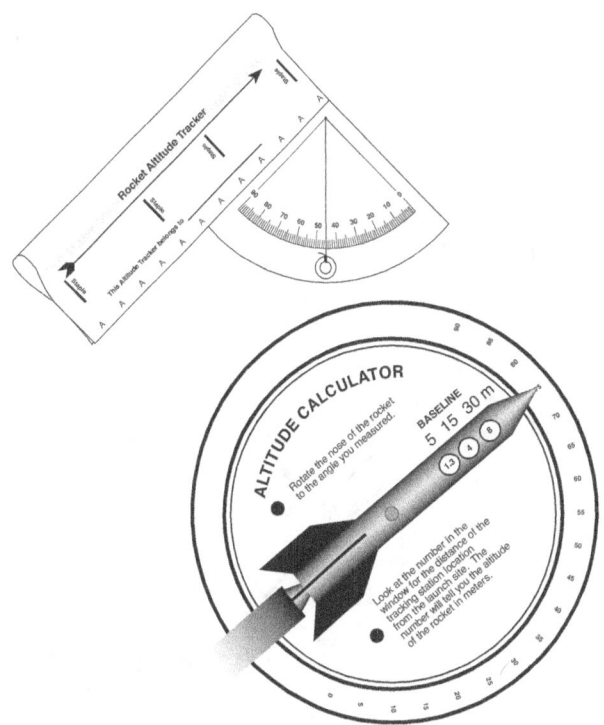

Management
Determining the altitude a rocket reaches in flight is a team activity. While one group of students prepares and launches a rocket, a second group measures the altitude the rocket reaches by estimating the angle of the rocket at its highest point from the tracking station. The angle is input into the altitude tracker calculator, and then the altitude is read. Roles are reversed so that everyone gets to launch and to track, depending upon the number of launches held and whether or not every student determines their own altitude.

The trackers and altitude calculators activity should take an hour or two. While waiting to launch rockets or track them, students can work on other projects.

Altitude tracking, as used in this activity, can be used with the Paper Rockets (page 43), 3-2-1 Pop! (page 58) and Bottle Rocket Assembly (page 85) activities and with commercial model rockets. The altitude calculator is calibrated for 16.4, 49.2 and 98.4 ft (5, 15 and 30 m) baselines. Use the 16.2 ft (5 m) baseline for Paper Rockets and 3-2-1 Pop rockets. Use the 49.2 ft (15 m) baseline for Project X-35, and use the 98.4 ft (30 m) baseline for launching commercial model rockets.

For practical reasons, the altitude calculator is designed for angles in increments of five degrees. Younger students may have difficulty in obtaining precise angle measurements with the altitude tracker. For simplicity's sake, round measurements off to the nearest five-degree increment and read the altitude reached directly from the altitude calculator. If desired, you can determine altitudes for angles in between the increments by adding

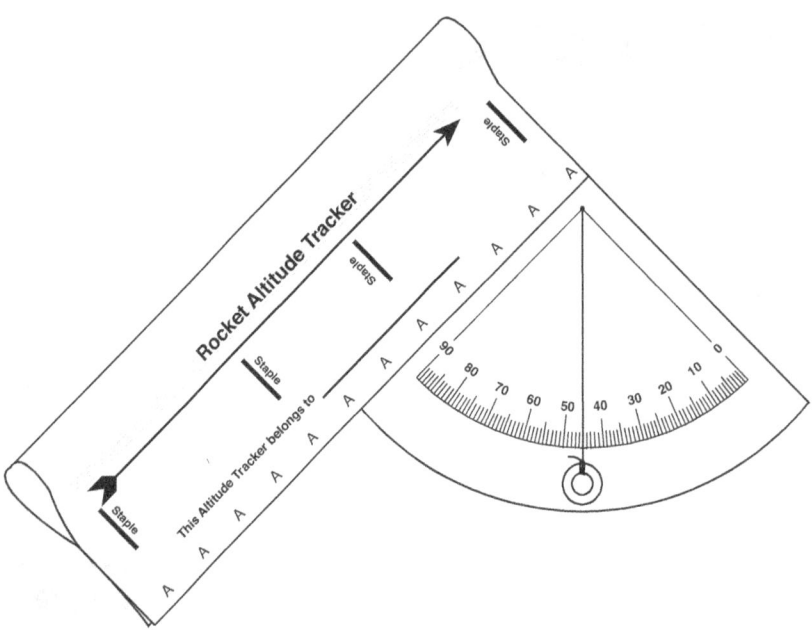

Completed Altitude Tracker Scope

the altitudes above and below the angle and dividing by 2. A more precise method for determining altitudes appears later in the procedures.

A teacher's aid or older student should cut out the three windows in the altitude calculator. A sharp knife or razor and a cutting surface work best for cutting out windows. The altitude tracker is simple enough for everyone to make one of their own, but they can also be shared. Students should practice taking angle measurements and using the calculator on objects of known height, such as a building or a flagpole, before calculating rocket altitude.

Background Information

This activity makes use of simple trigonometry to determine the altitude a rocket reaches in flight. The basic assumption of the activity is that the rocket travels straight up from the launch site. If the rocket flies away at an angle other than 90 degrees, the accuracy of the procedure diminishes. For example, if the rocket climbs over a tracking station, where the angle is measured, the altitude calculation will yield an answer higher than the actual altitude reached. On the other hand, if the rocket flies away from the station, the altitude measurement will be lower than the actual value. Tracking accuracy can be increased by using more than one tracking station to measure the rocket's altitude. Position a second or third station in different directions from the first station. Averaging the altitude measurements will reduce individual error.

Constructing the Altitude Tracker Scope Procedure

1. Copy the pattern for the altitude tracker on heavy weight paper.
2. Cut out the pattern on the dark outside lines.
3. Curl (do not fold) the B edge of the pattern to the back until it lines up with the A edge.

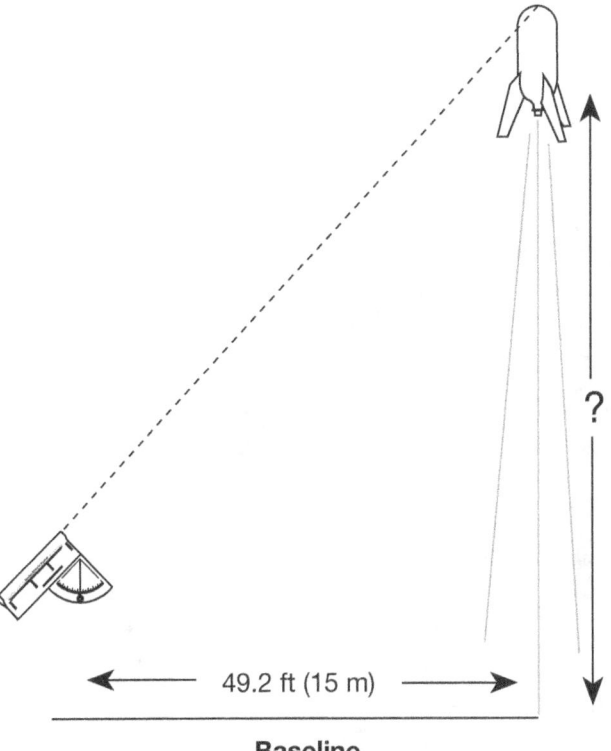

Baseline

and antacid-powered rockets, a 49.2 ft (15 m) distance is sufficient for bottle rockets, and a 98.4 ft (30 m) distance is sufficient for model rockets.)

2. As a rocket launches, the person doing the tracking will follow the flight with the sighting tube on the tracker. The tracker should be held like a pistol and kept at the same level as the rocket when it is launched. Continue to aim the tracker at the highest point the rocket reached in the sky. Have a second student read the angle the thread or string makes with the quadrant protractor. Record the angle.

Constructing the Altitude Calculator
Procedure

1. Copy the two patterns for the altitude calculator onto heavy weight paper or glue the patterns onto lightweight poster board. Cut out the patterns.
2. Place the top pattern on a cutting surface, and cut out the three windows.
3. Join the two patterns together where the center marks are located. Use a brass paper fastener to hold the pieces together. The pieces should rotate smoothly.

4. Staple the edges together where marked. If done correctly, the As and Bs will be on the outside of the tracker.
5. Punch a small hole through the apex of the protractor quadrant on the pattern.
6. Slip a thread or lightweight string through the hole. Knot the thread or string on the backside.
7. Complete the tracker by hanging a small washer from the other end of the thread as shown in the diagram above.

Using the Altitude Tracker
Procedure

1. Set up a tracking station location a short distance away from the rocket launch site, depending upon the expected altitude the rocket rose after launch. A vertical line is drawn to show the vertical distance of 16.4, 49.2 or 98.4 ft (5, 15 or 30 m) away. (Generally, a 16.4 ft (5 m) distance is sufficient for paper rockets

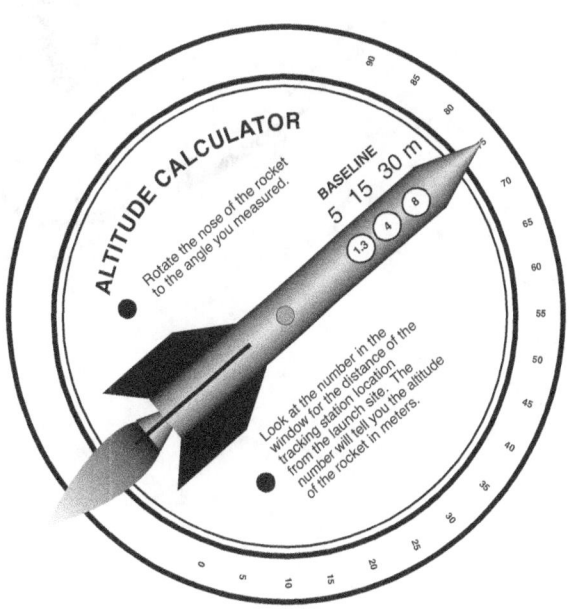

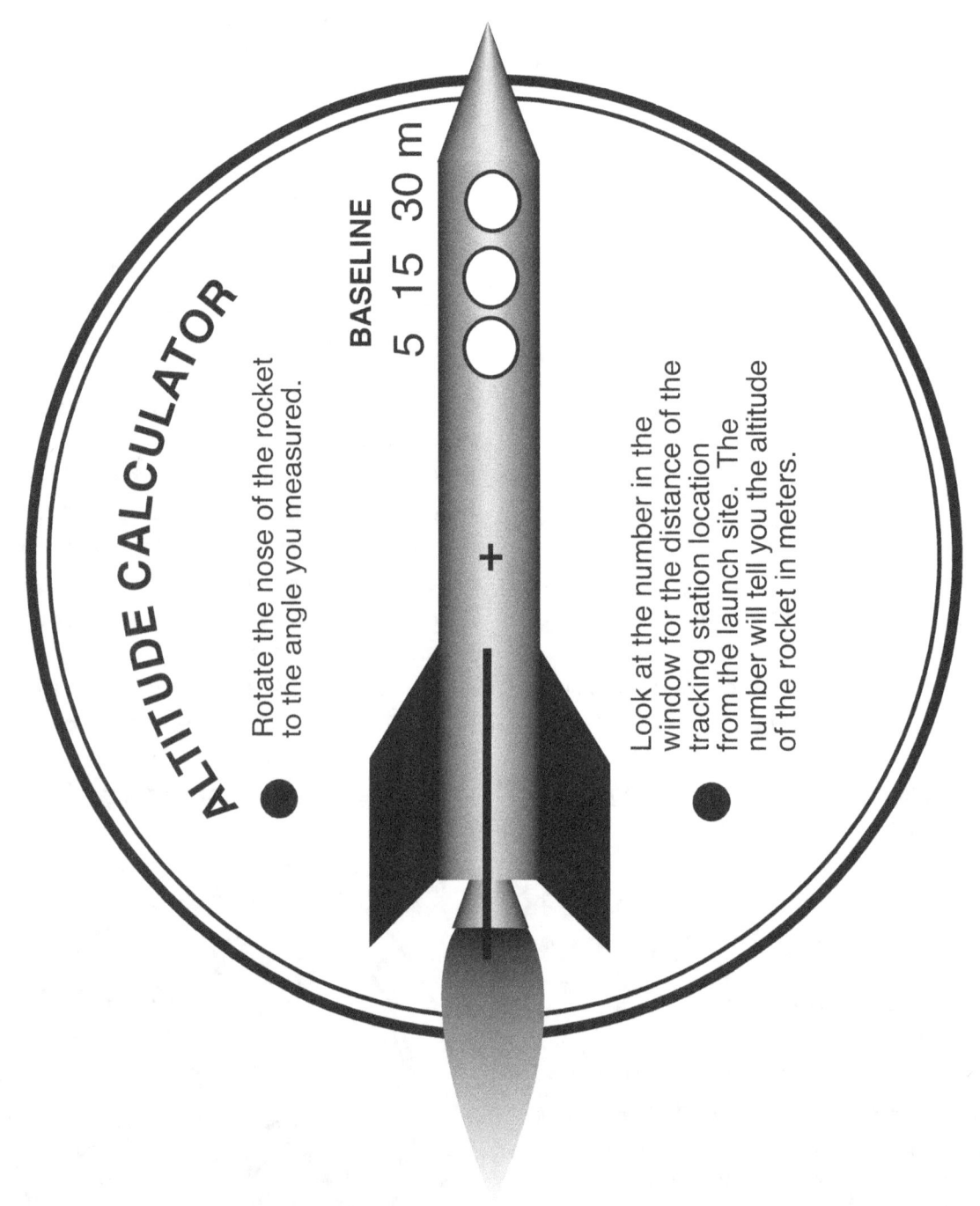

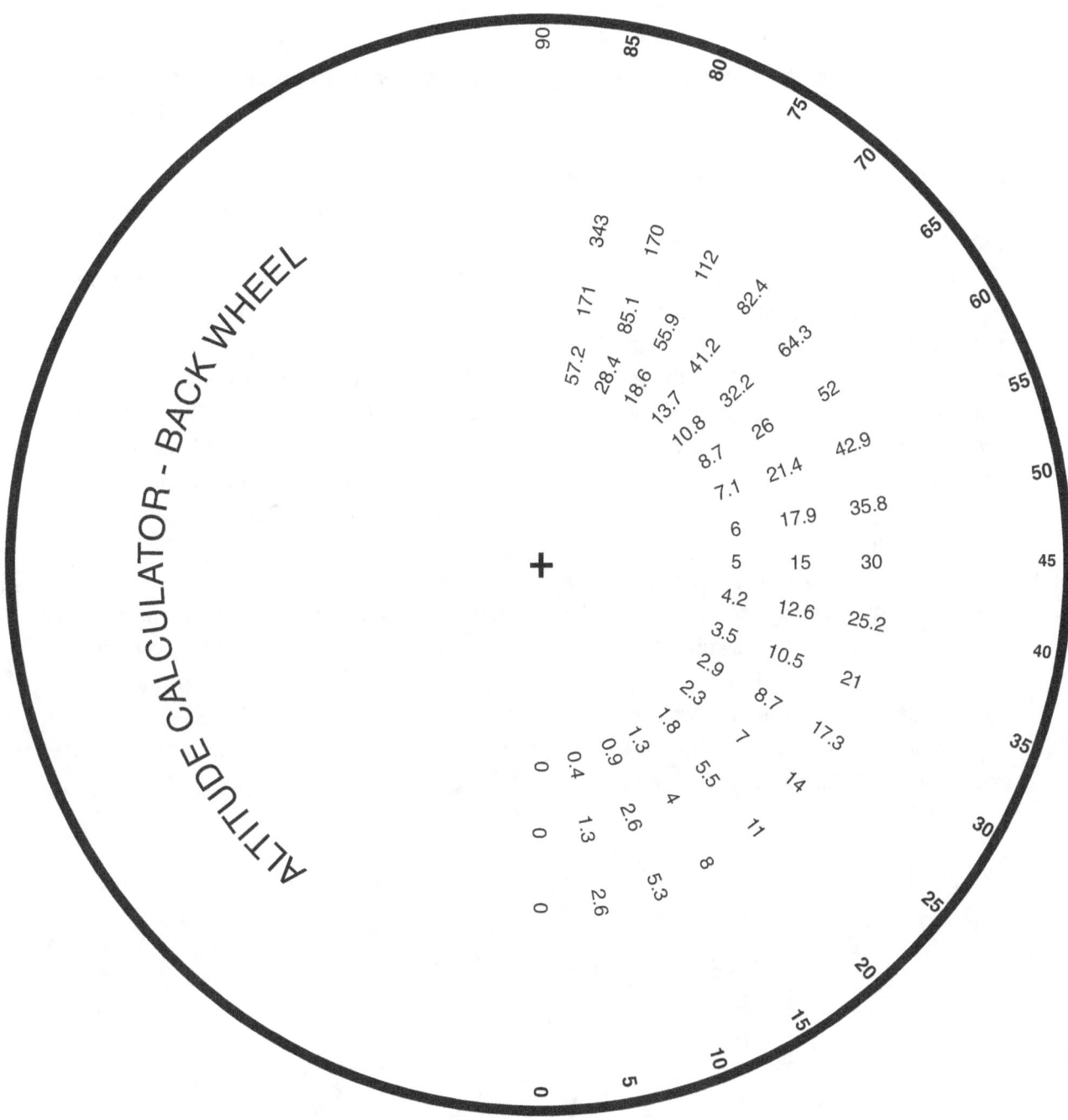

Determining the Altitude
Procedure
1. Use the altitude calculator to determine the height the rocket reached. To do so, rotate the inner wheel of the calculator so that the nose of the rocket pointer is aimed at the angle measured in step 2 of the previous procedure.
2. Read the altitude of the rocket by looking in the window. If you use a 5 m (16.5 ft) baseline, the altitude the rocket reached will be in the window beneath the 5. To achieve a more accurate measure, add the height of the person holding the tracker to calculate altitude. If the angle falls between two degree marks, average the altitude numbers above and below the marks.

Advanced Altitude Tracking
1. A more advanced altitude tracking scope can be constructed by replacing the rolled sighting tube with a fat milkshake straw. Use white glue to attach the straw along the 90-degree line of the protractor.
2. Once you determine the angle of the rocket, use the following equation to calculate altitude of the rocket:
Altitude = tan × baseline
3. Use a calculator with trigonometry functions to solve the problem or refer to the tangent table on page 116. For example, if the measured angle is 28 degrees and the baseline is 49.2 ft (15 m), the altitude is 26.15 ft (7.98 m).

$$\text{Altitude} = \tan 28° \times 49.2 \text{ ft } (15 \text{ m})$$

$$\text{Altitude} = 0.5317 \times 49.2 \text{ ft } (15 \text{ m}) = 26.15 \text{ ft } (7.98 \text{ m})$$

4. An additional improvement in accuracy can be obtained by using two tracking stations. Averaging the calculated altitude from the two stations will achieve greater accuracy. See the figure below.

Assessment
Have students demonstrate their proficiency with altitude tracking by sighting on a fixed object of known height and comparing their results. If employing two tracking stations, compare measurements from both stations.

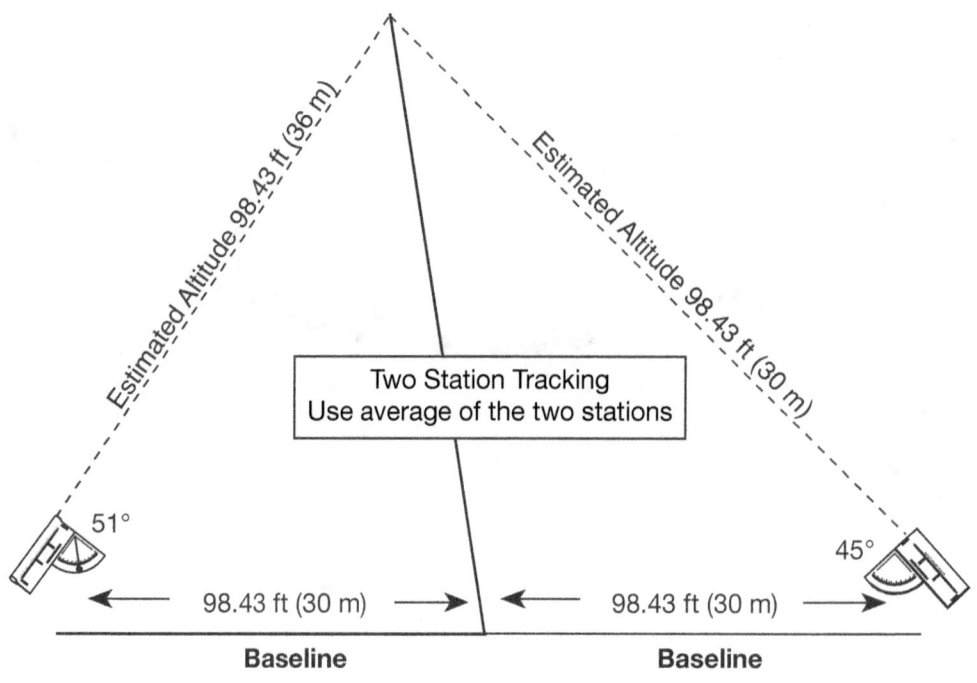

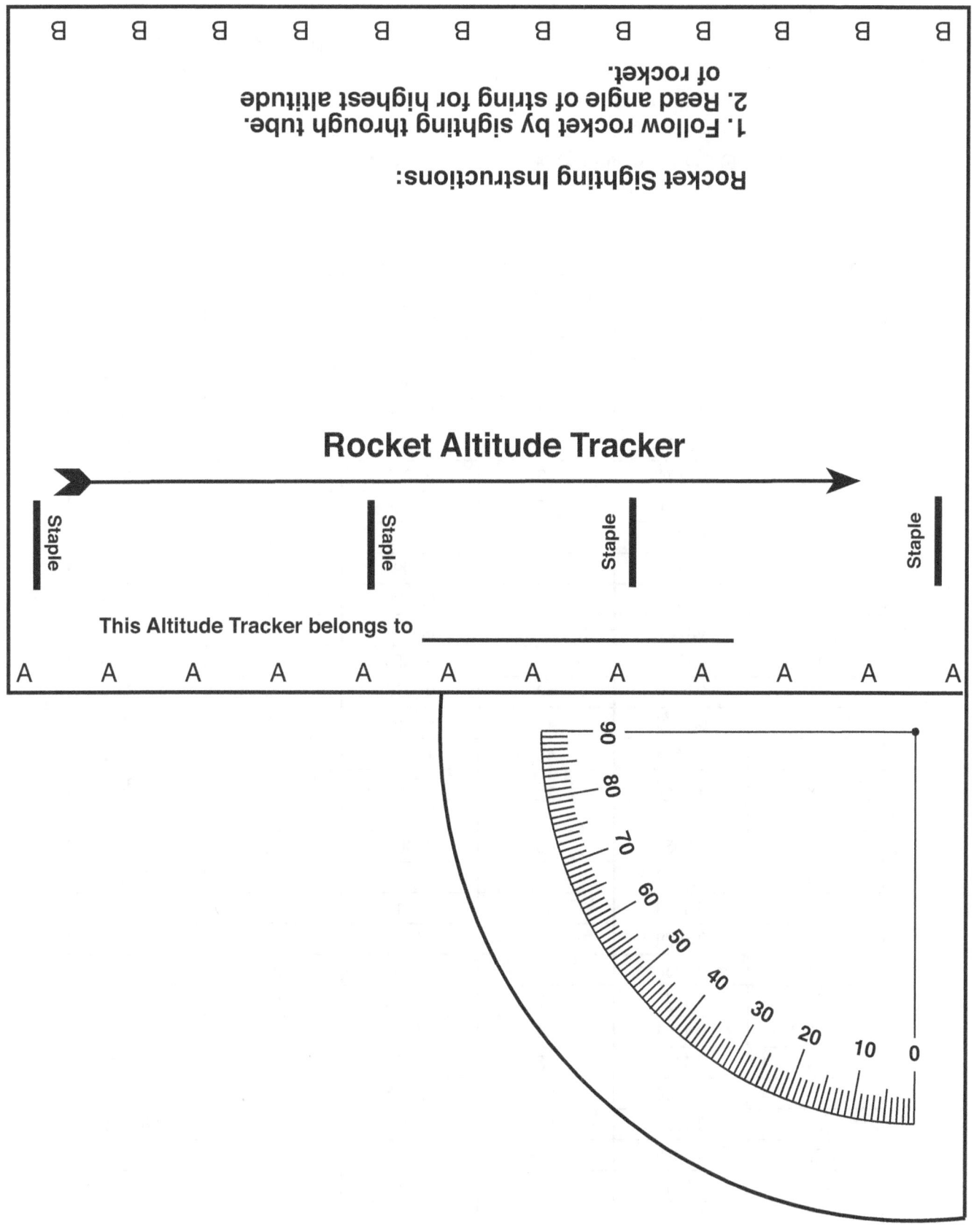

Extensions

- Why should the height of the person holding the tracker be added to the measurement of the rocket's altitude?
- Curriculum guides for model rocketry (available from model rocket supply companies) provide instructions for more sophisticated rocket tracking measurements. These activities involve two-station tracking with altitude and compass direction measurement and trigonometric functions.

Tangent table.

Degree	Tan	Degree	Tan	Degree	Tan
0	0.0000	31	0.6008	62	1.8807
1	0.0174	32	0.6248	63	1.9626
2	0.0349	33	0.6494	64	2.0603
3	0.0524	34	0.6745	65	2.1445
4	0.0699	35	0.7002	66	2.2460
5	0.0874	36	0.7265	67	2.3558
6	0.1051	37	0.7535	68	2.4750
7	0.1227	38	0.7812	69	2.6050
8	0.1405	39	0.8097	70	2.7474
9	0.1583	40	0.8390	71	2.9042
10	0.1763	41	0.8692	72	3.0776
11	0.1943	42	0.9004	73	3.2708
12	0.2125	43	0.9325	74	3.4874
13	0.2308	44	0.9656	75	3.7320
14	0.2493	45	1.0000	76	4.0107
15	0.2679	46	1.3055	77	4.3314
16	0.2867	47	1.0723	78	4.7046
17	0.3057	48	1.1106	79	5.1445
18	0.3249	49	1.1503	80	5.6712
19	0.3443	50	1.1917	81	6.3137
20	0.3639	51	1.2348	82	7.1153
21	0.3838	52	1.2799	83	8.1443
22	0.4040	53	1.3270	84	9.5143
23	0.4244	54	1.3763	85	11.3006
24	0.4452	55	1.4281	86	19.0811
25	0.4663	56	1.4825	87	19.0811
26	0.4877	57	1.5398	88	28.6362
27	0.5095	58	1.6003	89	57.2899
28	0.5317	59	1.6642	90	∞
29	0.5543	60	1.7320		
30	0.5773	61	1.8040		

The Scale of a Model Rocket

Objective
Students will show what scale is using dimensional proportions and ratios and how scales relate to models. Students will use unit conversions and compare the altitude of models scaled from the same design.

- Target concept: Altitude
- Preparation time: 30 min
- Duration of activity: 60 min
- Student group size: 3 to 15 students

Materials
- Dimension information sheet on Saturn V, Ares, Delta or other NASA rocket
- Large sheet of drawing paper or paper sheets taped together
- Pencils
- Ruler
- Long paper tubes, such as from wrapping paper or construction paper

Drawing Procedure
1. Take the linear dimensional data (length, width, etc.) from the fact sheet, and divide all the values by the scale factor of 100.
2. Using those scaled dimensions and the ruler, draw the rocket on the paper. All the proportions of length and width will be preserved in the shapes.

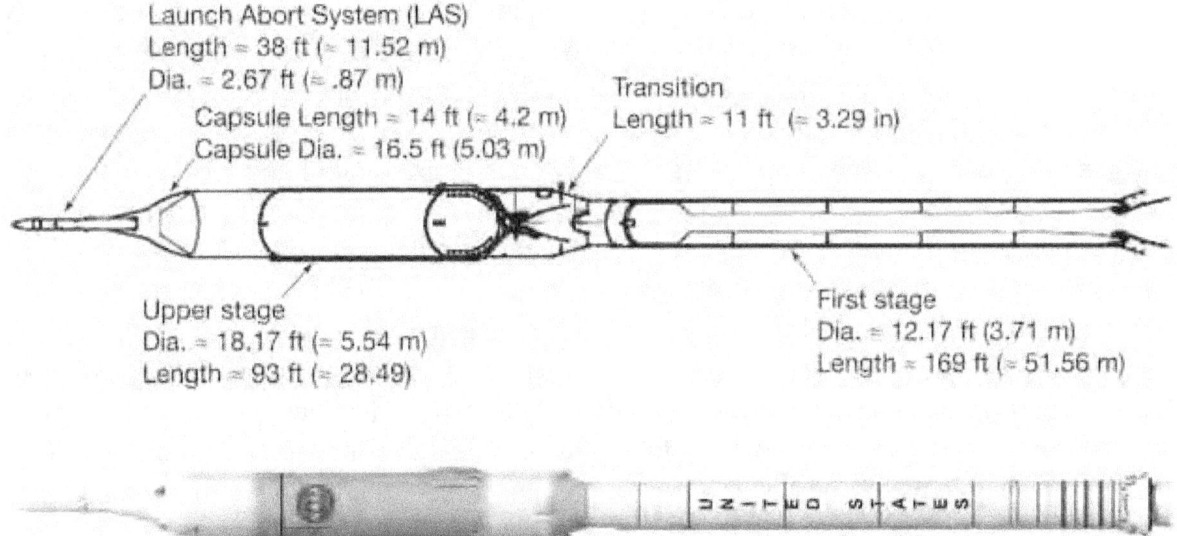

Ares I Crew Launch Vehicle Dimension Sheet
Total Vehicle Length 325 ft (99.1 m)

Launch Abort System (LAS)
Length ≈ 38 ft (≈ 11.52 m)
Dia. ≈ 2.67 ft (≈ .87 m)

Capsule Length ≈ 14 ft (≈ 4.2 m)
Capsule Dia. ≈ 16.5 ft (5.03 m)

Transition
Length ≈ 11 ft (≈ 3.29 in)

Upper stage
Dia. ≈ 18.17 ft (≈ 5.54 m)
Length ≈ 93 ft (≈ 28.49)

First stage
Dia. ≈ 12.17 ft (3.71 m)
Length ≈ 169 ft (≈ 51.56 m)

These are approximate lengths as indicated by the ≈.

Sample Work Sheet for Apollo Saturn

Rocket Parameter	Dimension		Scale (1/100)		Scale Dimension for Drawing	
	(ft)	(m)	(ft)	(m)	(ft)	(m)
Length	393.7	120	393.7/100	120/100	3.94	1.2
Diameter	32.8	10	32.8/100	10/100	0.33	0.1
Tail span	39.4	12	39.4/100	12/100	0.39	0.12
Tail length	9.8	3	9.8/100	3/100	0.10	0.03

Modeling procedure

1. Now determine another scale factor for multiplying the rocket dimensions to build a model. Select a paper tube about 1–3 in (2.54–7.62 cm) in diameter and measure it.
2. Find the scale proportion by dividing the diameter of the real rocket by the diameter of the tube to get a number like 1/222 or 1/72. This fraction is the scale factor.
3. Use this scale factor to multiply all the other rocket dimensions, and list the scaled values on your drawing. The first number to find will be the length of the tube.
4. Now build a model of the rocket according to those values around the size of the paper tube with other paper, tape and art supplies to make a nose cone and tail fins. This is just to be a display model of the physical proportions and not to be flown.

Questions

1. Get a scale model rocket kit and evaluate its closeness to scale of the real rocket by comparing its dimensions to your own calculations. How true to the scale factor is it in preserving the proportions? What other vehicles or objects are often seen in scale models?
2. If a Saturn V model is 3.6 ft (1.1 m) tall and 1/100 scale, how tall is the real rocket?
3. Scale can also be expressed in proportions, such as 1 in equals 1 ft. What would be the ratio for this scale?
4. This lesson looked at the scale of the dimensions. What other parameters of a rocket could be scaled in a model? What characteristics would have to be scaled for the model to fly like the full size vehicle?

Additional approach

The point of testing small engineering versions of a large device is to economically prove they work. This can be done because the scale model functions the same way as the full size version. To show this point, have students design their own rocket and build it to scale at two different sizes, such as one having a 0.75 in diameter body and the other a 1.5 in diameter body. One rocket will be small enough to fly on A-B size motors and the bigger one to fly on D-E motors or larger. Students should try to determine the drag value for the design first using a simulation software model for both design sizes. They will launch the small rocket and compare it to the simulation, then build and launch the large rocket and compare its altitude performance and drag value. Students will see the relationships of the computer model and the small flight-test model to the full-scale vehicle.

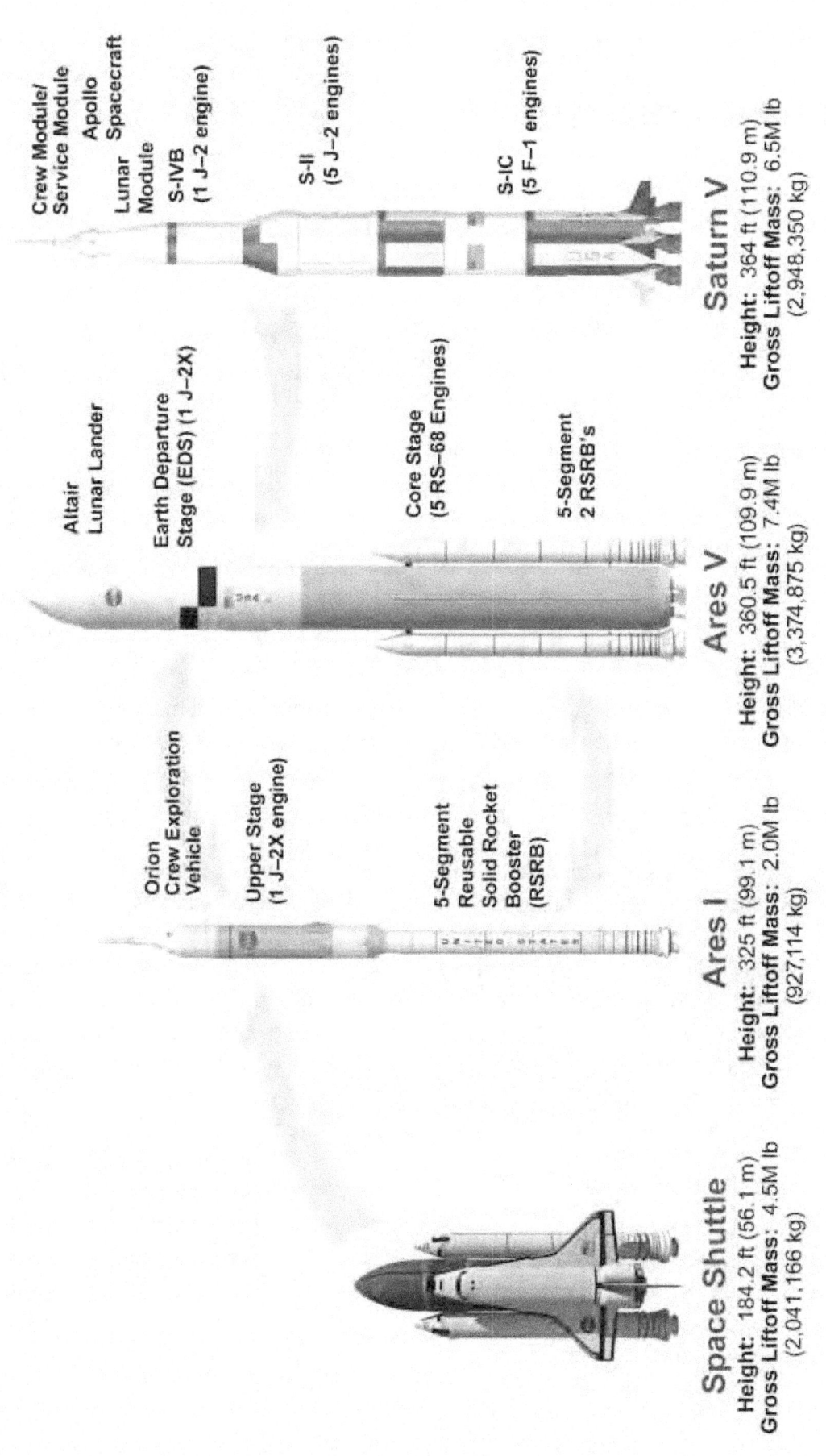

Rocket Motion Video Studies

Objective
Students will use a video camera to record motion, measure velocity and compute acceleration of a model rocket at launch. This lesson will teach measurement methodology, data interpretation and deriving acceleration from velocity.

- Target concept: Acceleration and velocity
- Preparation time: 60 min
- Duration of activity: 60 min to launch and 60 min to analyze
- Student group size: 3 to 12 students per adult

Materials
- Model rocket launch pad and firing system
- Small model rockets
- Motors
- Video camera on a tripod
- Computer for video editing or TV screen connected to camera
- Video editing software or camera that can view frame by frame
- Pole or PVC pipe about 10 ft (3.05 m) long
- Vinyl electrical tape
- Tape measure

Procedure
Launch and filming
1. For this exercise, a student is needed for each task to set up the pipe and pad, prep the rocket, launch the rocket, record the data, and operate the camera.

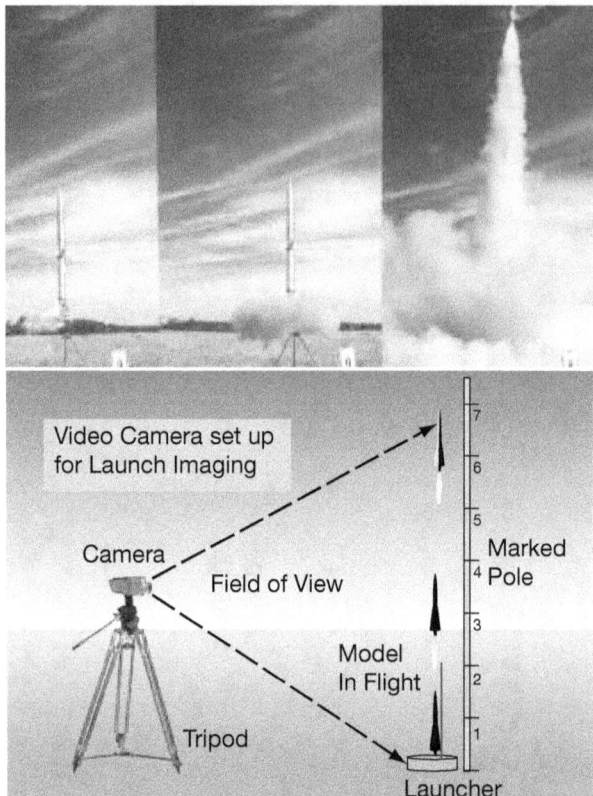

2. Measure and mark the pipe off in foot- (meter-) long increments. Put a wrap of tape around the pole at each foot (meter) mark (black tape on a white pipe for contrast). Number the marks so that the pipe becomes a giant ruler.
3. Set the pipe straight up in the launch field next to the launch pad. This can be done by driving a broomstick in the ground like a stake and setting the pipe on it. Set up the video camera so that the launcher and the entire pipe length is framed and focused within the viewfinder. The pipe should be directly behind the rocket.
4. Set up a small model on the pad to launch. Check the camera operation, and then set it to record. Launch the model. Stop the camera from recording. Repeat for several other launches, and note which motors are flown in each scene. A data sheet will help listing flight order, motor type, camera settings and display counter index.

Launch and filming data sheet

Flight	Motor	Counter Index of First Motion	Count at First Mark	Count at Second Mark

Video Analysis
1. Connect the camera to the computer to analyze your recording.
2. Note in the video where the nose of the rocket or the fins are against the markings on the pipe. Step through the frames until you see the moment the rocket begins to move.
3. Count the number of frames from that point to where the nose (or fin) crosses the next mark on the pipe.
4. From that point, count again the frames for the nose to reach the next mark, and so on until the rocket is out of view. You might like to save and print the best frames and work from a hard copy.

Calculations
Check the camera manual to find the video frame speed, which is likely 30 frames per second. This means each frame is 1/30 of a second. The rocket speed is $v = y/t$ where y is the distance moved, and t is time.

Calculate the rocket velocity from the actual displacement in time.

For example, if the rocket takes five frames to move up 1 ft (0.3 m), then five frames (5 x 1/30 sec) is 1/6 sec, and 1 ft ÷ 1/6 s = 6 ft/s (0.3 m ÷ 1/6 s = 1.8 m/s).

If the rocket moves 3-ft (0.91-m) marks between one frame, then
3 ft ÷ 1/30 s = 90 ft/s (0.91 m ÷ 1/30 s = 27.3 m/s).

To find the acceleration, first compute a final velocity for the last frames at the top of the pipe. Then, count up the frames of the entire flight from first motion to the last view of the rocket to get a total time. Acceleration will be $a = v/t$.

If the final velocity is 90 ft/s (27.3 m/s) over a total time of 1/3 second, then

a = 90 ft/s ÷ 1/3 s = 270 ft/s^2
(27.3 m/s ÷ 1/3 s = 81.9 m/s^2).

Compare these values to the predicted values of the same rocket in a simulation program.

Questions
1. How does the thrust value given in the motor code affect the observed acceleration; that is, how does the take off of the same rocket with a B4-4 motor compare to its launch with a B6-4 motor?
2. What shutter speeds on the camera work best to freeze the motion of rocket flights?
3. Does changing the contrast of the image during analysis help the resolution or clarity?

Additional approach
Get a small strobe light that brightly flashes at a known constant rate that can be easily seen in daylight. This may be a bicycle hazard light or a toy blinker. Use a clear plastic nose section to contain the strobe in a rocket. Use an appropriately sized rocket and motor to

safely carry the weight of the strobe. Test the strobe with the video camera by moving the strobe through the field of view and noting how the flashes appear in the image. Adjust the camera exposure to be sure the flashes appear bright enough to be clearly recorded.

Set up the pipe again by the launch pad, but position the camera further back or zoom out so that it sees space about five times the pipe length above the pad. Launch the rocket, and record the flight with the strobe light flashing.

On the video editor, find the frames where the flashes appear and use them as time markers. Count the frames between the peaks of the flashes. Approximate the distance moved by the rocket by referring to the pipe length. Using the known time of the flashes, find the velocity by dividing the approximated distance traveled by the rocket by the time.

For another approach, back up the camera far enough to see the entire rocket flight in the viewfinder. The increasing spacing of the flashes along the flight path illustrates the rocket's acceleration.

Rocket Parachute Flight Duration Study

Part 1: Flight Measurements

Objective
To understand the flight time performance effects and trades and to relate time to velocity, students will measure the flight time of rockets of various sizes and parachutes types, but using the same motor type.

- Target concept: Velocity
- Preparation time: 60 min
- Duration of activity: 60 min
- Student group size: 3 to 12 students per adult

Materials
- Several identical and some assorted size rockets
- Various standard size parachutes to test
- Plastic shopping bags
- Thread or string for making custom size parachutes
- B6-4 motors, or type A motors if needed for small fields
- Stopwatches
- Data sheet

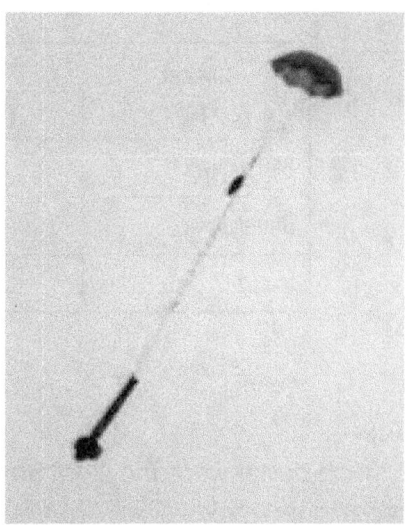

Procedure
1. For this exercise, a student is needed for each task to set up the pad, launch the rocket, time and record the data. Each student can have their own rocket or work in teams of three.
2. Tell the students to design a system (they will select a rocket and parachute) in order to achieve the longest total flight time on a B motor. All rockets must fly on the same type motor.
3. Prepare and launch the rockets and record each flight time with the rocket information on the data sheet. The flight time for the stopwatch operation is from the moment of first motion by the rocket off the pad until any part of it touches the ground. Two timers are recommended for better accuracy.
4. After the first round of launches, the students will review the data sheet of all flights. Discuss which rockets had the highest and lowest flight times and why.
5. The students then experiment with their rocket selection and parachute types to determine how to improve and extend their flight time. They can design and make their own parachute. They can investigate packing techniques, shroud line length and other chute design parameters.
6. Students will then launch another round of rockets with an adjusted and improved design, following the same timing process, and observe and record the results.

Sample data sheet.

Rocket Type	Chute Diameter (in)	Chute Diameter (cm)	Flight Time (s)
Starsquawk	12	30.48	33
Starsquawk	18	45.72	55
Icarus	24	60.96	56

Questions
1. How much better were the flight times of the redesigned rockets?
2. How does the flight time of a large rocket with a large parachute compare to a small rocket with a small parachute?
3. What is the main variable affecting descent time?

Part 2: Model Rocket Parachute Size Calculation

Objective
The size of the parachute and weight of the rocket determine how fast it will all come down. This equation shows the relation of all the rocket parameters to velocity. Use it to select the right size parachute for a rocket.

Procedure
Determine the size for a parachute for a typical rocket with the following equation:

$$D = \sqrt{8mg / \pi \rho C_d v^2}$$

where:
- D is the chute diameter in feet or meters.
- m is the rocket mass in pounds or kilograms.
- g is the acceleration of gravity = 32 ft/s² (9.8 m/s²).
- π is 3.14159
- ρ is the density of air = 0.076 lb/ft³ (1.22 kg/m³).
- C_d is the drag coefficient of the chute, which is 0.75 for a parasheet (flat plastic sheet used for a model rocket parachute
- v is the desired velocity at landing (9.8 ft/s (3 m/s) or less)

D is the parachute diameter you want to find. The rocket mass m can be found from the manufacturer's catalog if it is a standard kit, or you can weigh it on a scale. All the other values are constants in this case and given above.

For example, if m = 2.198 oz = 0.137 lb (62.3 g = 0.0623 kg). Inputting all the values and multiplying them out yields

$$D = 12 \left(\sqrt{(8 \cdot 0.137 \cdot 32) / (3.14 \cdot 0.076 \cdot 9.8^2)} \right) = 17 \text{ in}$$

or $\sqrt{(8 \cdot 0.0623 \cdot 9.81) / (3.14 \cdot 1.22 \cdot 3^2)} = 0.435$ m.

Standard rocket parachutes are often made to have an 18-in (0.457-m) diameter.

Parachute diameters.

Rocket Type	Measured Chute Diameter		Total Weight		Calculated Diameter	
	(in)	(m)	(oz)	(kg)	(in)	(m)
Starsquawk	12	0.305	4	0.113	23	0.584
Icarus	24	0.670	6	0.170	28	0.711

Calculations

Using the weight of the rockets flown in part 1 and the given values in the parachute equation, solve for the parachute diameter.

Questions

1. How does this equation apply to the previous exercise in understanding parachute flight performance?
2. What's the most critical parameter in the equation?
3. How did the parachute size determined by the equation compare to the diameter of the actual parachute used in the flights?

Predicting and Measuring Rocket Parachute Drift Rate

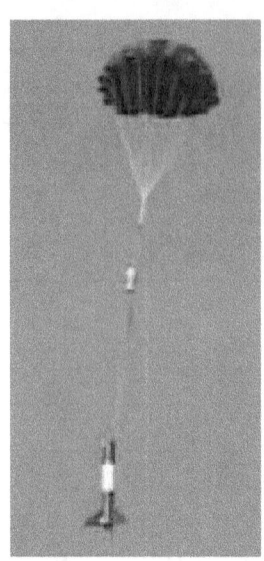

Objective
Students will use vector velocity calculations to determine the vertical descent rate and horizontal drift distance of a rocket parachute. Predicted values will be compared to measured data. Students will learn to manage and interpret data and evaluate results.

- Target concept: Velocity
- Preparation time: 60 min
- Duration of activity: 60 min
- Student group size: 3 to 12 students per adult

Materials
Same as for the parachute duration experiment, but including the following:

- Ping-pong anemometer to measure wind speed (Plans to make one are in Appendix A.)
- Scale to measure rocket weight
- 100 ft (30 m) long tape measure
- Data sheet

Procedure
1. This is similar to the previous parachute duration experiment. Referring to that basic procedure, launch the rocket, but in this case, measure the time of the flight from the moment when the parachute opens to the time of landing.
2. Measure the wind speed with the anemometer near and about the same time as the flight.
3. Use the tape or a long string to measure the distance from the launch pad to where the rocket lands. You can pace this off and approximate the distance if a straight-line measurement is not possible.
4. Weigh the landed rocket with its burnt-out motor. Record all of this data on the sheet.

Calculations
Calculating Parachute Descent Velocity
The descent velocity equation for the rocket can be determined by rearranging the equation for parachute size in the previous section to get velocity. Use the following equation to calculate the vertical descent velocity from

Sample data sheet sample.

Rocket Type	Weight		Chute Diameter		Flight Time (s)	Landing Distance		Wind Speed	
	(oz)	(kg)	(in)	(m)		(ft)	(m)	(ft/s)	(m/s)
Icarus	6.2	0.176	24	0.61	56	356	108.51	6	1.83

the known rocket weight and parachute diameter,

$$v = \sqrt{\frac{8mg}{\pi \rho C_d D^2}}.$$

Input the values for a rocket and the constants previously given into this equation and solve for velocity.

Calculating the drift distance

The horizontal drift velocity of the rocket will be assumed to be the same as the wind velocity. From the measured wind velocity and flight time, calculate the horizontal drift distance of the parachute from the following equation:

$$\text{Drift} = \text{Time}_{chute} \times V_{wind}$$

where
- Drift = How far the rocket will travel from the launch site
- V_{wind} = Wind velocity
- Time_{chute} = Parachute time in seconds

Compare this value to the drift distance you measured with the tape.

Example

If you have a wind blowing with a V = 12.43 mi/hr (20 km/hr) and convert it to 18.23 ft/s (5.557 m/s) and the parachute flight was 33 seconds, the drift distance will be,

Drift = 33 s × 18.23 ft/s (33 s × 5.557 m/s) = 601.61 ft (183.38 m).

Questions

1. What is the highest wind speed you can fly a rocket in before its parachute will likely drift from your pad to outside the clear area of your field?
2. What is the maximum altitude a rocket can have that will still land in the field? You will need to estimate your field dimensions to get an answer.

Additional approach

Suppose the predicted altitude y of a rocket is 1,000 ft (304.8 m). Given the weight and parachute size, determine the descent velocity v. Now use $t=y/v$ to calculate the parachute descent time t. That time can now be multiplied with the wind velocity to find the drift distance. Taken all together, these values can predict the landing point of a rocket or where a launching pad needs to be placed in a field.

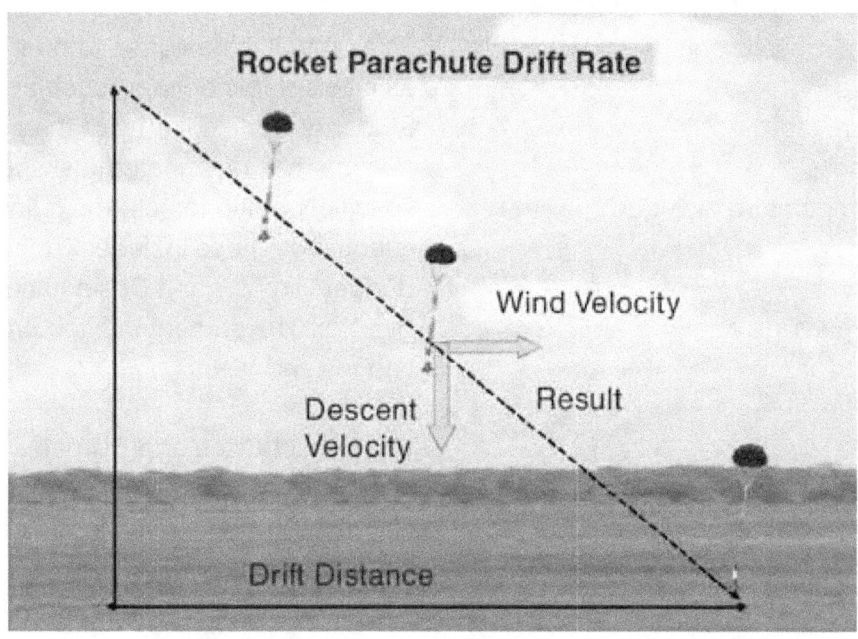

Rocket Parachute Drift Rate

Project Enterprise

Objective
To demonstrate rocketry principles through a cooperative, problem solving simulation.

- Target concept: Altitude, velocity and acceleration
- Preparation time: 1-hour duration of activity
- Activity time: 10 1-hour periods
- Student group size: Teams of two to four students

Description
Teams simulate the development of a commercial proposal to design, build and launch a rocket.

Materials and Tools
(All supplies need to be available per group)
- 2-liter soda bottles
- 1-liter soda bottles
- Film canisters
- Aluminum soda cans
- Scrap cardboard and poster board
- Large cardboard panels
- Duct tape
- Electrical tape
- Glue sticks
- Low-temperature glue gun
- Water
- Clay
- Plastic garbage bags
- Crepe paper
- String
- Paint
- Safety glasses
- Bottle rocket launcher (see page 88)
- Altitude calculator (see page 110)

Management
Prior to this project, students should have the opportunity to design, construct and launch a bottle rocket, evaluating various water volumes and air pressures and calculating the altitude traveled by these rockets. See Water Bottle Rocket Assembly page 85 and Altitude Tracking page 109.

This project is designed to offer students an interdisciplinary approach to life skills. Students work in teams of threes. Each member has designated tasks for their specific job title to help the team function effectively. These include: Project Manager, Budget Director and Design and Launch Director. The student section provides badges and tasks.

The project takes approximately two weeks to complete and includes a daily schedule of tasks. Students may need additional time to complete daily tasks.

Collect all building materials and copy all reproducible items before beginning the activity. Be sure to make several copies of the order forms and checks for each group.

Allow enough time on the first day for the students to read and discuss all sheets and determine how they apply to the project schedule. Focus on the student score sheet to allow a clear understanding of the criteria used for assessment of the project.

Background Information
This project provides students with an exciting activity to discover practical demonstrations of force and motion in actual experiments, while dealing with budgetary restraints and deadlines reflected in real life situations.

The students should have a clear understanding of rocket principles dealing with Newton's Laws of Motion found on page 159 and Practical Rocketry found on page 7 before beginning this project.

Procedure
Refer to the student sheets. The events for day three and day six call for teacher demonstrations on how to make nose cones and how to determine the center of mass and the center of pressure. See page 138 for an explanation.

Assessment
Assessment will be based on documentation of three designated areas: each group's project journal, silhouette and launch results. See Student Score Sheet for details.

Request for Proposals

The United Space Authority (USA) is seeking competitive bids for a new advanced rocket launch vehicle that will reduce the costs of launching payloads into Earth orbit. Interested companies are invited to submit proposals to USA for designing and building a rocket that will meet the following criteria.

The objectives of Project Enterprise are:

1. Design and draw a bottle rocket plan to scale (1 square = 1 in [2.54 cm]).
2. Develop a budget for the project and stay within the budget allowed.
3. Build a test rocket using the budget and plans developed by your team.
4. Identify rocket specifications and evaluate rocket stability by determining center of mass and center of pressure and conducting a swing test. See page 139 for an explanation.
5. Display fully illustrated rocket design in class. Include dimensional information, location of center of mass and center of pressure, and flight information, such as time aloft and altitude reached.
6. Successfully test launch rocket achieving maximum vertical distance and accuracy.
7. Successfully and accurately complete rocket journal.
8. Develop a cost analysis and demonstrate the most economically efficient launch.

Proposal Deadline
2 weeks

Project Enterprise Schedule

Project Enterprise Schedule: Day 1
- Form rocket companies.
- Brainstorm ideas for design and budget.
- Sketch preliminary rocket design.

Project Enterprise Schedule: Day 2
- Develop materials and budget list.
- Develop scale drawing.

Project Enterprise Schedule: Day 3
- Demonstration: nose cone construction.
- Issue materials and begin construction.

Project Enterprise Schedule: Day 4
- Continue construction.

Project Enterprise Schedule: Day 5
- Complete construction.

Project Enterprise Schedule: Day 6
- Demonstration: Find center of mass and center of pressure.
- Introduce rocket silhouette construction and begin rocket analysis.

Project Enterprise Schedule: Day 7
- Finish silhouette construction and complete prelaunch analysis.
- Hang silhouette.
- Perform swing test.

Project Enterprise Schedule: Day 8
- Launch Day!

Project Enterprise Schedule: Day 9
- Complete post launch results and silhouette documentation.
- Prepare journal for collection.
- Documentation and journal due at beginning of class tomorrow.

Project Enterprise Checklist

Project Grading
- Documentation (See Project Journal below. Must be complete, neat, accurate and on time.): 50 percent.
- Proper display and documentation of rocket silhouette: 25 percent.
- Launch data (measurements, accuracy and completeness): 25 percent.

Project Awards
USA will award exploration contracts to the companies with the top three rocket designs based on the above criteria. The awards are valued at the following:

- First: $10,000,000
- Second: $5,000,000
- Third: $3,000,000

Project Journal
Check off the following items as you complete them:
1. Creative cover with member's names, date, project number and company name.
2. Certificate of Assumed Name (Name of your business).
3. Scale drawing of rocket plans. Clearly indicate scale. Label: Top, Side and End View.
4. Budget Projection.
5. Balance Sheet.
6. Canceled checks (staple or tape checks in ascending numerical order, four to a sheet of paper).
7. Prelaunch Analysis.
8. Rocket Launch Day Log.
9. Score Sheet (Part 3).

Badges

Each group member will be assigned specific tasks to help their team function successfully. All team members assist with design, construction, launch and paper work. Enlarge the badges and glue them front and back to poster board. Cut out the slot and attach a string.

X-35 Project Manager

Oversees the project. Keeps others on the task. Only person who can communicate with teacher.

- Arrange all canceled checks in ascending numerical order.
- Make a neat copy of the team's Rocket Journal.
- Use appropriate labels as necessary.
- Check over balance sheet. List all materials used in rocket construction.
- Complete silhouette information and display properly in room.
- Assist other team members as needed.

X-35 Design and Launch Director

Supervises design and construction of rocket. Directs others during launch.

- Make a neat copy of the Launch Day Log.
- Use appropriate labels as necessary.
- Arrange to have a creative cover made.
- Assist other team members as needed.

X-35 Budget Director $

Keeps accurate account of money expenses and pay bills. Must sign all checks.

- Arrange all canceled checks in order and staple four to a sheet of paper.
- Check over budget projection sheet. Be sure to show total project cost estimates.
- Check over balance sheet. Be sure columns are complete and indicate a positive or negative balance.
- Complete part three of the score sheet.
- Assist other team members as needed.

Badge Front   **Badge Back**

Rockets: An Educator's Guide with Activities in Science, Mathematics, and Technology   EG-2003-01-108-HQ

133

Project Enterprise Budget

Each team will be given a budget of $1 million. Use the money wisely and keep accurate records of expenditures. Once your money runs out, you will operate in the "red," and this will count against your team score. If you are broke at the time of launch, you will be unable to purchase rocket fuel. You will then be forced to launch with compressed air. You may only purchase as much rocket fuel as you can afford at the time of launch.

All materials not purchased from listed subcontractors will be assessed an import duty tax of 20 percent of the market value. Materials not on the subcontractors list will be assessed an Originality Tax of $5,000 per item.

A project delay penalty fee will be assessed for not working, lacking materials, etc. This penalty fee could be as high as $300,000 per day.

Approved Subcontractor List

Subcontractor	Material		Market Price
Bottle Engine Corporation	2-L bottle		$200,000
	1-L bottle		$150,000
Aluminum Cans Ltd	Can		$ 50,000
International Paper Corporation	Cardboard	1 sheet	$ 25,000
	Tag board	1 sheet	$ 30,000
	Manila paper	1 sheet	$ 40,000
	Silhouette panel	1 sheet	$100,000
International Tape and Glue Company	Duct tape	50-cm segments	$ 50,000
	Electrical tape	100-cm segments	$ 50,000
	Glue stick	1 ea	$ 20,000
Aqua Rocket Fuel Service	1 ml		$ 300
Strings, Inc	1 m		$ 5,000
Plastic Sheet goods	1 bag		$ 5,000
Common Earth Corporation	Modeling clay	100 g	$ 5,000
NASA Launch Port	Launch		$100,000
NASA Consultation	Question	1 ea	$ 1,000

Project Enterprise Balance Sheet

Company Name _____

Check No.	Date	To	Amount ($)	Balance ($)

Rocket Measurements For Scale Drawing

Project No. _____
Date _____

Company Name _____

Use American Standard measurements to measure and record the data in the blanks below. Be sure to accurately measure all objects that are constant (such as the bottles) and those you will control (like the size and design of fins). If additional data lines are needed, use the back of the sheet.

Data table.

Object	Length (ft)	Width (in)	Diameter (in)	Circumference (in)

Using graph paper, draw a side, top and bottom view or your rocket, to scale (1 square = 1 in), based on the measurements as recorded above. Attach your drawings to this paper.

Scale Drawing

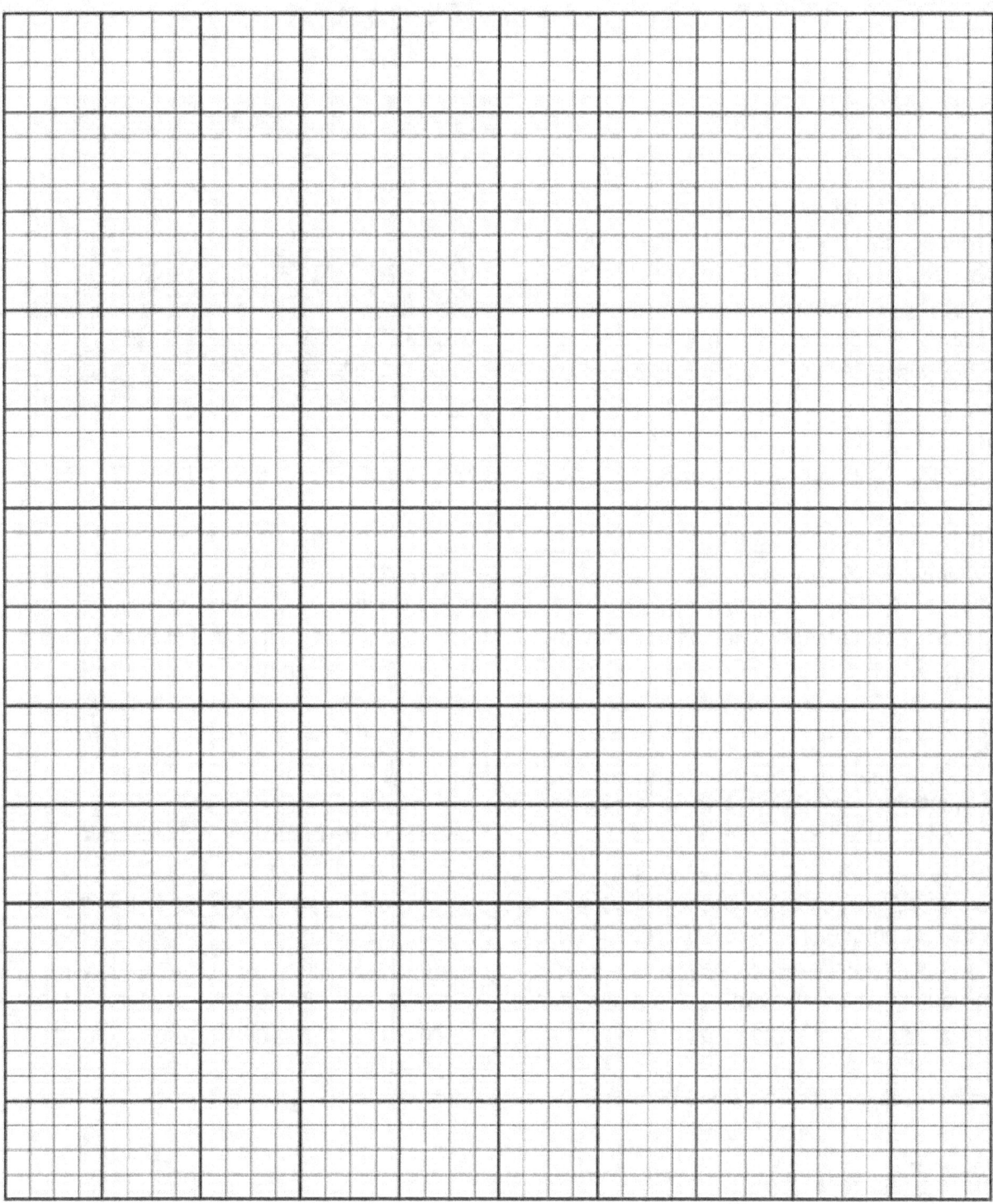

Note: One square equals 1 in.

Rocket Stability Determination

A rocket that flies straight through the air is said to be a stable rocket. A rocket that veers off course or tumbles wildly is said to be an unstable rocket. The difference between the flight of a stable rocket and an unstable rocket depends upon its design. All rockets have two distinct centers. The first is the center of mass (CM). This is a point about which the rocket balances. If you could place a ruler edge under this point, the rocket would balance horizontally like a seesaw. What this means is that half of the mass of the rocket is on one side of the ruler edge, and half is on the other. CM is important to a rocket's design because if a rocket is unstable, the rocket will tumble about this center.

The other center in a rocket is the center of pressure (CP). This is a point where half of the surface area of a rocket is on one side and half is on the other. The CP differs from the CM in that its location is not affected by the placement of payloads in the rocket. This is just a point based on the surface of the rocket, not what is inside. During flight, the pressure of air rushing past the rocket will balance half on one side of this point and half on the other. You can determine the CP by cutting out an exact silhouette of the rocket from cardboard and balancing it on a ruler edge.

The positioning of the CM and the CP on a rocket is critical to its stability. The CM should be towards the rocket's nose, and the CP should be towards the rocket's tail for the rocket to fly straight. That is because the

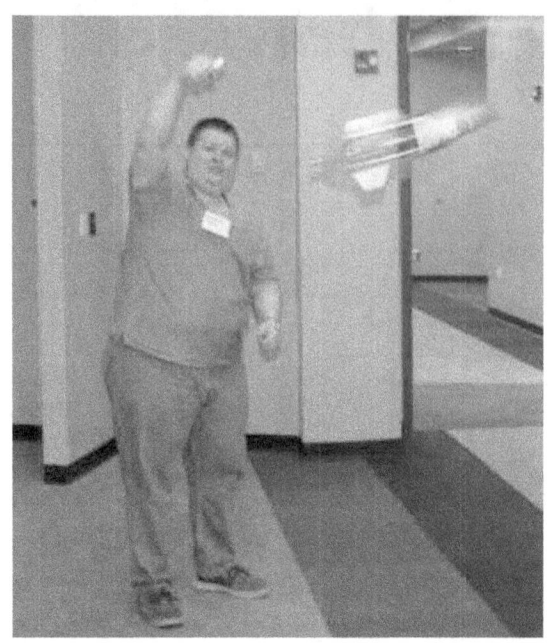

lower end of the rocket (starting with the CM and going downward) has more surface area than the upper end (starting with the CM and going upward). When the rocket flies, more air pressure exists on the lower end of the rocket than on the upper end. Air pressure will keep the lower end down and the upper end up. If the CM and the CP are in the same place, neither end of the rocket will point upward. The rocket will be unstable and tumble.

Stability Determination Instructions

1. Tie a string loop around the middle of your rocket. Tie a second string to the first so that you can pick it up. Slide the string loop to a position where the rocket balances. You may have to temporarily tape the nose cone in place to keep it from falling off.

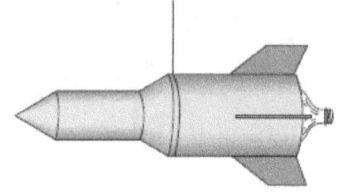

2. Draw a straight line across the scale diagram of the rocket you made earlier to show where the ruler's position is. Mark the middle of the line with a dot. This is the rocket's CM.
3. Lay your rocket on a piece of cardboard. Carefully trace the rocket on the cardboard, and cut it out.
4. Lay the cardboard silhouette you just cut out on the ruler and balance it.
5. Draw a straight line across the diagram of your rocket where the ruler is. Mark the middle of this line with a dot. This is the CP of the rocket.

If your CM is in front of the CP, your rocket should be stable. Proceed to the swing test. If the two centers are next to or on top of each other, add more clay to the nose cone of the rocket. This will move the CM forward. Repeat steps 2 and 3, and then proceed to the swing test.

Swing Test
1. Tape the string loop you tied around your rocket in the previous set of instructions so that it does not slip.
2. While standing in an open place, slowly begin swinging your rocket in a circle. If the rocket points in the direction you are swinging it, the rocket is stable. If not, add more clay to the rocket nose cone or replace the rocket fins with larger ones. Repeat the stability determination instructions, and then repeat the swing test.

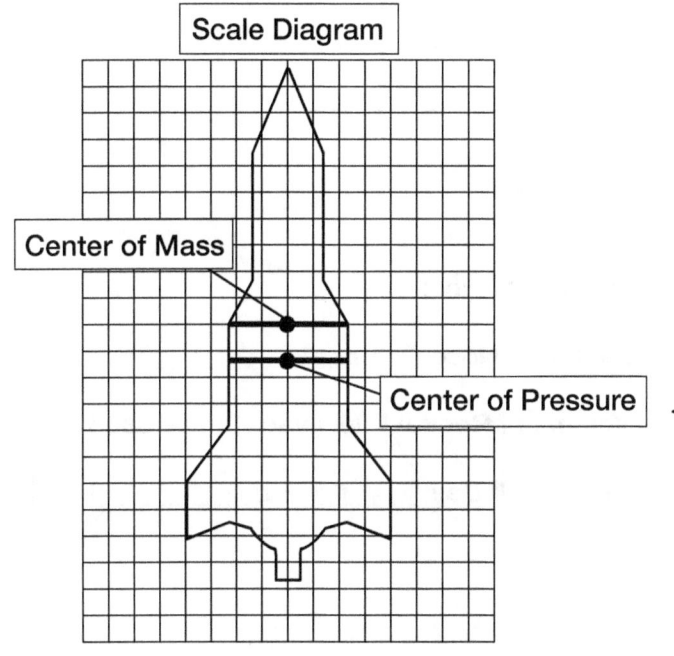

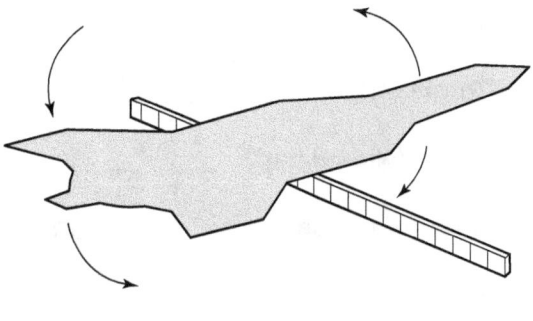

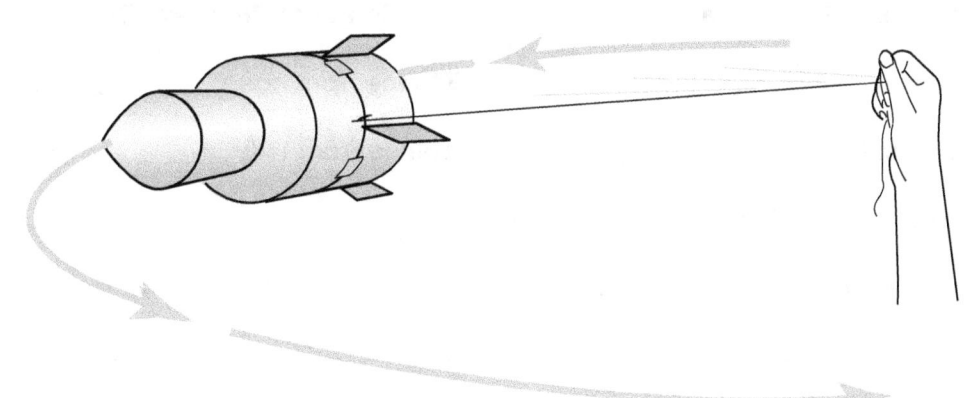

Prelaunch Analysis

Prelaunch Analysis

Company Name: _____ Project Number: ☐

Employee Name: _____

Job Title: _____

Employee Name: _____

Job Title: _____

Employee Name: _____

Job Title: _____

Rocket Specifications

Total Mass: _____oz   Number of Fins: _____

Total Length: _____in   Length of Nose Cone: _____in

Width (widest part): _____in   Volume of Rocket Fuel (water) to be used

Circumference: _____in   on Launch Day: _____oz _____qt

Rocket Stability

Center of Mass (CM)	Center of Pressure (CP)
Distance from nose: _____in	Distance from nose: _____in
Distance from tail: _____in	Distance from tail: _____in

Distance of CM from SP: _____in

Did your rocket pass the swing test? _____

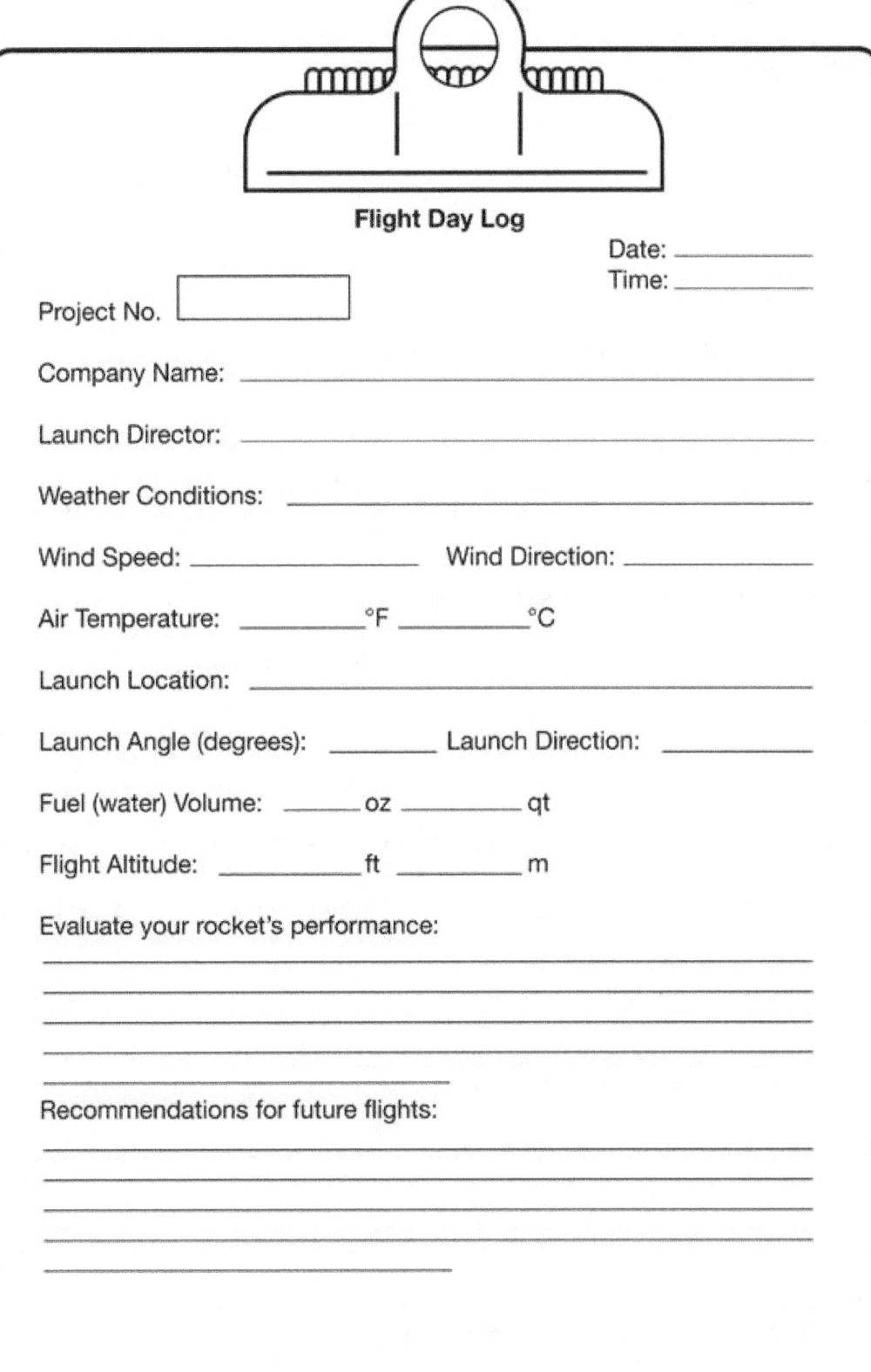

Project Enterprise Score Sheet

Total Score: _____ **Project No.** _____

Date: _____

Company Name: _____

Part I: Documentation (50 percent of project grade)

    Neatness: _____ Completeness: _____

    Accuracy: _____ Order: _____

    On Time: _____ Score: _____

Part II: Silhouette (25 percent of project grade)

    Neatness: _____ Completeness: _____

    Accuracy: _____ Order: _____

    On Time: _____ Score: _____

Part III: Launch Results (25 percent of project grade) (teams complete this section)

    a. Rocket Altitude: _____ Rank: _____

    b. Expenditures and Penalty Fees: _____
       (Check total from Balance Sheet)

    c. Final Balance: _____
       (New balance on Balance Sheet)

    d. Efficiency (Cost/Foot [Cost/Meter]): _____
       (Divide investment [b] by rocket altitude [a])

    e. Contract Award: _____

    f. Profit: _____
       (Contract award [f] minus expenditures [b])

                                             Score: _____

Glossary

Acceleration: Increase in speed or velocity, a change in velocity

Action: A force (push or pull) acting on an object. See Reaction.

Active controls: Devices on a rocket that move to control the rocket's direction in flight.

Airfoil: A streamlined shape given to fins or wings for maximum aerodynamic efficiency in flight.

Altitude: Extent or distance upward, height

Apogee: The highest point of a rocket's flight path.

Attitude control rockets: Small rockets that are used as active controls to change the attitude (direction) a rocket or spacecraft is facing in outer space.

Burnout: The point at which propellant is exhausted in a motor.

Canards: Small movable fins located towards the nose cone of a rocket.

Case: The body of a solid propellant rocket that holds the propellant.

Center of mass (CM): The point in an object about which the object's mass is centered.

Center of pressure (CP): The point in an object about which the object's surface area is centered.

Chamber: A cavity inside a rocket where propellants burn.

Combustion chamber: See Chamber.

Drag: Friction forces in the atmosphere that drag on a rocket to slow its flight.

Escape velocity: The velocity an object must reach to escape the pull of Earth's gravity.

Extravehicular activity (EVA): Spacewalking.

Fins: Arrow-like wings at the lower end of a rocket that stabilize the rocket in flight.

Friction: The interaction of the surface of one body against that of another causing a slowing of motion.

Fuel: The chemical that combines with an oxidizer to burn and produce thrust.

Gimbaled nozzles: Tiltable rocket nozzles used for active controls.

Igniter: A device that ignites a rocket's engine(s).

Injectors: Showerhead-like devices that spray fuel and oxidizer into the combustion chamber of a liquid-propellant rocket.

Insulation: A coating that protects the case and nozzle of a rocket from intense heat.

Liquid propellant: Rocket propellants in liquid form.

Mass: The amount of matter contained within an object.

Mass fraction (MF): The mass of propellants in a rocket divided by the rocket's total mass.

Microgravity: An environment that imparts to an object a net acceleration that is small compared to that produced by Earth at its surface.

Motion: Movement of an object in relation to its surroundings.

Movable fins: Rocket fins that can move to stabilize a rocket's flight.

Newton's First Law of Motion: A body remains at rest or in motion with a constant velocity unless acted upon by an external force.

Newton's Second Law of Motion: A force on an object will cause the object to accelerate in the direction of the force. The object will accelerate directly proportional to the force and inversely proportional to the mass of the object. Force = Mass × Acceleration.

Newton's Third Law of Motion: For every action, there is an equal and opposite reaction.

Nose cone: The cone-shaped front end of a rocket.

Nozzle: A bell-shaped opening at the lower end of a rocket through which a stream of hot gases is directed.

Oxidizer: A chemical containing oxygen compounds that permits rocket fuel to burn both in the atmosphere and in the vacuum of space.

Passive Controls: Stationary devices, such as fixed rocket fins, that stabilize a rocket in flight.

Payload: The cargo (scientific instruments, satellites, spacecraft, etc.) carried by a rocket.

Propellant: A mixture of fuel and oxidizer that burns to produce rocket thrust.

Pumps: Machinery that moves liquid fuel and oxidizer to the combustion chamber of a rocket.

Reaction: A movement in the opposite direction from the imposition of an action. See Action.

Rest: The absence of movement of an object in relation to its surroundings.

Regenerative cooling: Using the low temperature of a liquid fuel to cool a rocket nozzle.

Solid propellant: Rocket fuel and oxidizer in solid form.

Stages: Two or more rockets stacked on top of each other in order to reach higher altitudes or have a greater payload capacity.

Throat: The narrow opening of a rocket nozzle.

Terminal velocity: The constant speed obtained by a falling object when the upward drag on the object balances the downward force of gravity.

Unbalanced force: A force that is not countered by another force in the opposite direction.

Velocity: A moving object's speed and direction of movement.

Vernier rockets: Small rockets that use their thrust to help direct a larger rocket in flight.

Appendix A – Assembly Instructions

Straw Rocket Launcher Building Instructions

Objective
Build an air pulse (stomp) launcher to propel straw and paper rockets used in the activities of this guidebook.

Preparation Time
10 min

Materials
- 33.8-oz (1-L) plastic soft drink bottle
- Drinking straw or similar size tube. (A straight tube will work, but a straw or tube with a right angle elbow bend in it is better.)
- Modeling clay

Assembly
Insert the straw or tube into the mouth of the bottle about an inch. Seal the mouth around the straw with the modeling clay to completely fill the bottle's neck. The length of the outside tube should be as long as the rocket to allow it to slide on completely.

Operation
- Set the bottle on its side on a table or solid surface.
- Slide a straw rocket all the way onto the tube. It should slide freely.
- Point the rocket in a safe direction away from people.
- To launch the rocket, squeeze the bottle quickly or strike down on the side of it with your fist.
- Restore the roundness of the bottle by removing crumples or creases before the next launch.

Water Bottle Rocket Launcher

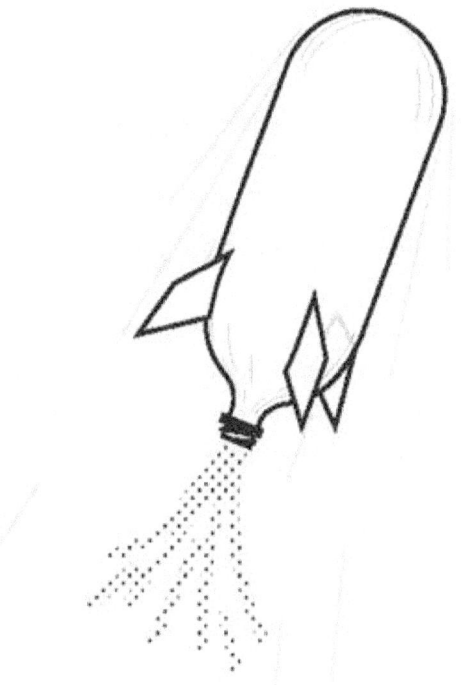

Objective
To construct a bottle rocket launcher for use with the Bottle Rocket and Project Enterprise activities.

Description
Students construct a bottle launcher from off-the-shelf hardware and wood using simple tools.

Management
Consult the materials and tools list to determine what you will need to construct a single bottle rocket launcher. The launcher is simple and inexpensive to construct. Air pressure is provided by means of a hand-operated bicycle pump. The pump should have a pressure gauge for accurate comparisons between launches. Most needed parts are available from hardware stores. In addition, you will need a tire valve from an auto parts store and a rubber bottle stopper from a school science experiment. The most difficult task is to drill a 0.375 inch hole in the mending plate called for in the materials list. An electric drill is a common household tool. If you do not have access to one, or do not wish to drill the holes in the metal mending plate, find someone who can do the job for you. Ask a teacher or student in your school's industrial arts shop, a fellow teacher or the parent of one of your students to help.

If you have each student construct a bottle rocket, having more than one launcher may be advisable. Because the rockets are projectiles, safely using more than one launcher will require careful planning and possibly additional supervision. Please refer to the launch safety instructions.

Materials and Tools
- Four 5 in corner irons with 12.75 in wood screws to fit
- One 5 in mounting plate
- Two 6 in spikes
- Two 10 in spikes or metal tent stakes
- Two 5 in by 0.25 in carriage bolts with six 0.25 in nuts
- One 3 in eyebolt with two nuts and washers
- 0.75 in diameter washers to fit bolts
- One number 3 rubber stopper with a single hole
- One Snap-in Tubeless Tire Valve (small 0.453 in hole, 2 in long)
- Wood board 12 × 18 × 0.75 in
- One 2-liter plastic bottle
- Electric drill and bits including a 0.375 in bit
- Screwdriver
- Pliers or open-end wrench to fit nuts
- Vice
- 12 ft of 0.25 in cord
- Pencil
- Bicycle pump with pressure gauge

Background Information

Like a balloon, air pressurizes the bottle rocket. When released from the launch platform, air escapes the bottle, providing an action force accompanied by an equal and opposite reaction force (Newton's Third Law of Motion). Increasing the pressure inside the bottle rocket produces greater thrust since a large quantity of air inside the bottle escapes with a higher acceleration (Newton's Second Law of Motion). Adding a small amount of water to the bottle increases the action force. The water expels from the bottle before the air does, turning the bottle rocket into a bigger version of a water rocket toy available in toy stores.

Construction Instructions

1. Prepare the rubber stopper by enlarging the hole with a drill. Grip the stopper lightly with a vice and gently enlarge the hole with a 0.375 in bit and electric drill. The rubber will stretch during cutting, making the finished hole somewhat less than 0.375 in.
2. Remove the stopper from the vice and push the needle valve end of the tire stem through the stopper from the narrow end to the wide end.
3. Prepare the mounting plate by drilling a 0.375 in hole through the center of the plate. Hold the plate with a vice during drilling and put on eye protection. Enlarge the holes at the opposite ends of the plates, using a drill bit slightly larger than the holes to do this. The holes must be large enough to pass the carriage bolts through them. (See Attachment of Mending Plate and Stopper diagram on page 149.)

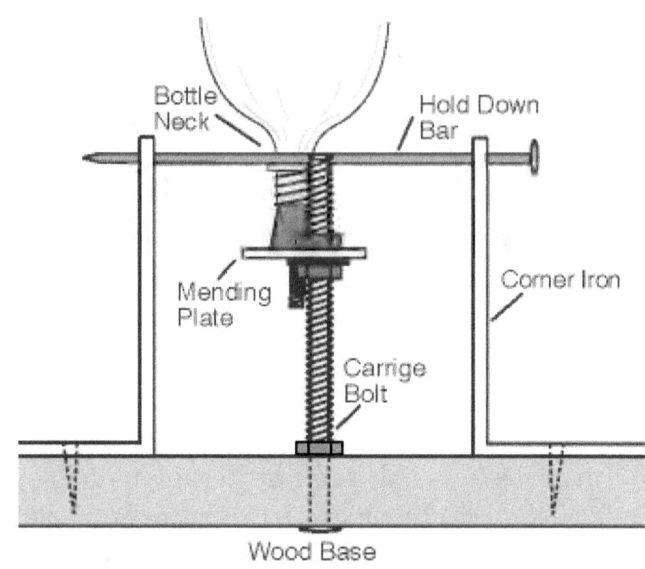

Attachment of Mending Plate and Stopper

Positioning Corner Irons

4. Lay the mending plate in the center of the wood base and mark the centers of the two outside holes that you enlarged. Drill holes through the wood big enough to pass the carriage bolts through.
5. Push and twist the tire stem into the hole you drilled in the center of the mounting plate. The fat end of the stopper should rest on the plate.
6. Insert the carriage bolts through the wood base from the bottom up. Place a hex nut over each bolt and tighten the nut so that the bolt head pulls into the wood.
7. Screw a second nut over each bolt and spin it about half way down the bolt. Place a washer over each nut, and then slip the mounting plate over the two bolts.
8. Press the neck of a 2-liter plastic bottle over the stopper. You will be using the bottle's wide neck lip for measuring in the next step.

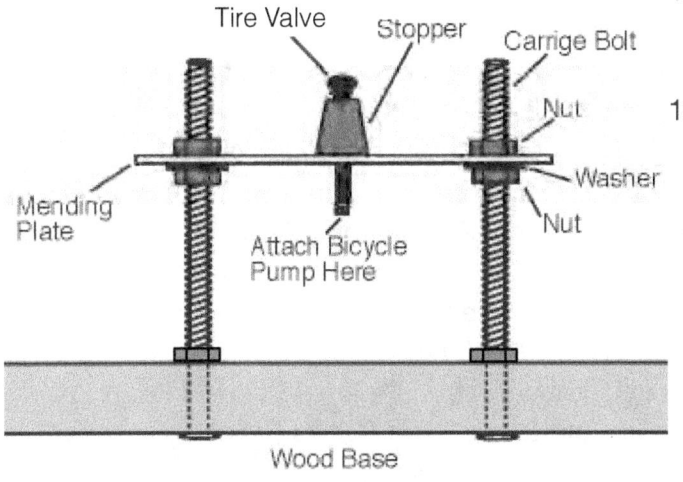

Positioning Corner Irons

9. Set up two corner irons so they look like book ends (see page 148). Insert a spike through the top hole of each iron. Slide the irons near the bottle's neck so that the spike rests immediately above the wide neck lip. The spike will hold the bottle in place while you pump up the rocket. If the bottle is too low, adjust the nuts beneath the mounting plate on both sides to raise it.

10. Set up the other two corner irons as you did in the previous step (see page 150). Place them on the opposite side of the bottle. When you have the irons aligned so that the spikes rest above and hold the bottle lip, mark the centers of the holes on the wood base. For more precise screwing, drill small pilot holes for each screw, and then screw the corner irons tightly to the base.

11. Install an eyebolt to the edge of the opposite holes for the hold down spikes. Drill a hole, and hold the bolt in place with washers and nuts on top and bottom (see page 150).

12. Attach the launch pull cord to the head end of each spike. Run the cord through the eyebolt.

13. Make final adjustments to the launcher by attaching the pump to the tire stem and pumping up the bottle. Refer to the launching instructions for safety notes. If the air seeps out around the stopper, the stopper is too loose. Use a pair of pliers or a wrench to raise each side of the mounting plate in turn to press the stopper with slightly more force to the bottle's neck. When satisfied with the position, thread the remaining hex nuts over the mounting plate and tighten them to hold the plate in position.

14. Drill two holes through the wood base along one side (see page 150). The holes should be large enough to pass large spikes or metal tent stakes. When the launch pad is set up on a grassy field, the stakes will hold the launcher in place when you yank the pull cord. The launcher is now complete (see page 150).

Launch Safety Instructions

1. Select a grassy field that measures approximately 100 ft (30.48 m) across. Place the launcher in the center of the field, and anchor it in place with the spikes or tent stakes.
   Note: If it is a windy day, place the launcher closer to the side of the field from which the wind is coming so that the rocket will drift onto the field as it comes down.

2. Have each student or student group set up their rocket on the launch pad. Other students should stand back several feet (meters). It will be easier to keep observers away by roping off the launch site.

3. After the rocket is attached to the launcher, the student pumping the rocket should put on eye protection. The rocket should be pumped no higher than about 50 lb/in^2 (8.93 kg/cm^2) of pressure.
4. When pressurization is complete, all students should stand in back of the rope for the countdown.
5. Before conducting the countdown, be sure the place where the rocket is expected to come down is clear of people. Launch the rocket when the recovery range is clear.
6. Only permit the students launching the rocket to retrieve it.

Extensions

Look up the following references for additional bottle rocket plans and other teaching strategies:

- Hawthorne, M.; and Saunders, G.: "It's Launchtime!," *Science and Children*, Volume 30 No. 5, pp. 17–19, 39, 1993.
- Rogis, J.: "Soaring with Aviation Activities," *Science Scope*, Volume 15 No. 2, pp. 14–17, 1991.
- Winemiller, J.; Pedersen, J.; and Bonnstetter, R.: "The Rocket Project," *Science Scope*, Volume 15 No. 2, pp. 18–22, 1991.

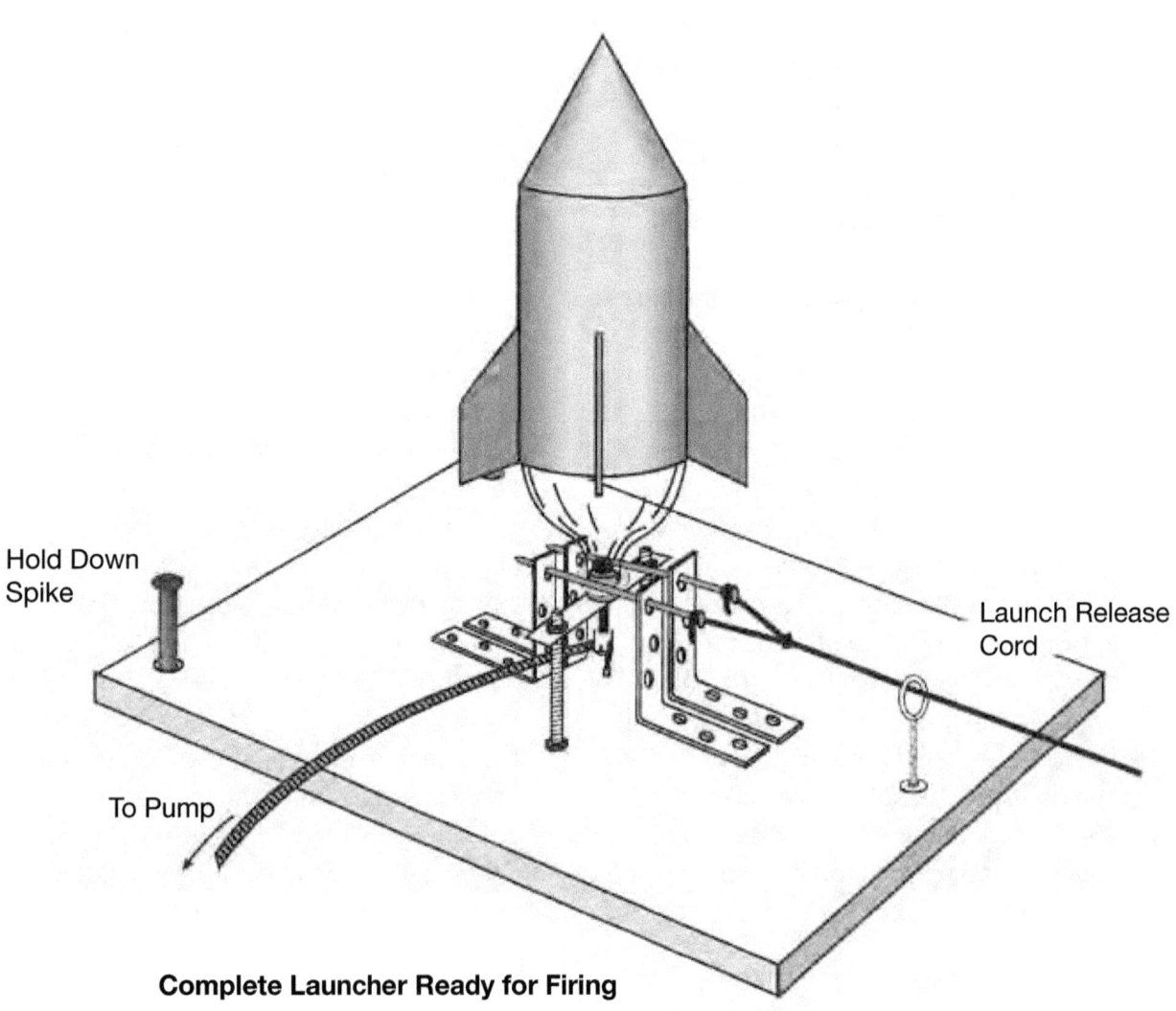

Complete Launcher Ready for Firing

The Ping-Pong Ball Pendulum Anemometer

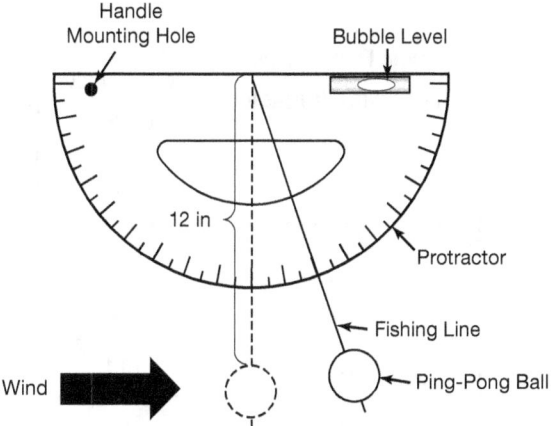

Objective
This is how to make a simple device to measure wind speed. It can be used to understand meteorology, but is applied in this guidebook to learn about rocket parachute drift speed. A pendulum anemometer operates by the wind blowing a bob suspended from an angular scale. The angle the bob is displaced from neutral corresponds to the wind speed.

Materials
- One 180 degree, 6 in clear plastic protractor
- One standard ping-pong ball
- One lightweight, monofilament line or strong thread
- One small bubble level
- One 0.5 in diameter wooden dowel handle or piece of broomstick
- Long thumbtack
- Glue
- Needle or small sharp knife point
- Marker
- Drill with a 0.0625 in bit

Assembly
1. Pierce the ping-pong ball to make a small hole. Thread one end of the line through the ball with the needle. Anchor the line with a dab of glue, but do not use any kind that might dissolve the ping-pong plastic. If this is difficult, just use a small piece of good strong tape to hold the string on the ball.
2. Next, glue the level to the protractor, parallel to the baseline. Be sure you align it correctly.
3. Pass the other end of the line through the index hole of the protractor. Adjust the length so that the top of the ball is 12 in (30.48 cm) from the suspension point at the hole. Then tie off or glue the line in place. If the line is a light color, you may want to darken it with a pen or marker where it crosses the scale for easier reading.
4. Drill a 0.0625 in hole in the protractor on the top edge on the side opposite the level. The 12 in (30.48 cm) length of handle can be cut from a broomstick or dowel to have a flat end. Attach the protractor to the handle by pressing the thumbtack through the hole in the protractor into the end of the wood stick. See finished anemometer on page 152.
5. Use a marker to label the protractor with wind scale values as follows:
   a. Mark the 90° mark of the protractor 0 mph.
   b. 80° mark 8 mph.
   c. 70° mark 12 mph.
   d. 60° mark 15 mph.
   e. 50° mark 18 mph.
   f. 40° mark 21 mph.
   g. 30° mark 26 mph.
   h. 20° mark 33 mph.

If you prefer to not mark the protractor, just read the protractor's angle graduations and find the wind speed from this fine scale chart.

Protractor scale chart.

Angle Fishing Line Makes with Protractor (degrees)	Wind Speed (mph)
90	0.0
85	5.8
80	8.2
75	10.1
70	11.8
65	13.4
60	14.9
55	16.4
50	18.6
45	19.5
40	21.3
35	23.4
30	25.7
25	28.6
20	32.4

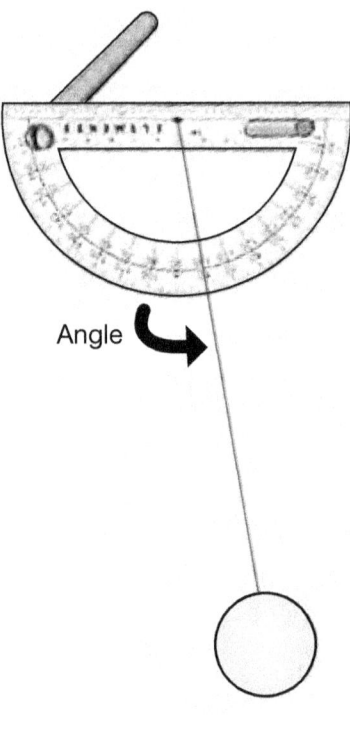

Operation

1. Take your reading in an open area.
2. Hold the handle out with the wind on your side and point the anemometer into the wind so that the ball moves the line along the protractor. It should be held away from the body to avoid the turbulent airflow surrounding you.
3. Sight on the level and turn your handle to adjust the tilt of the protractor and center the bubble. Then just read the mark where the line crosses the scale. You may want a helper to hold the anemometer level while you take the reading. Just keep your body at right angles to the wind.
4. Since not every degree in the chart is given as a speed, you may have to interpolate between the values. As with many instruments, the line may oscillate around the true value, so you have to watch for the maximum and minimum swings to average a reading. Gusts and lulls will bounce the ball around too, but with practice you can read the anemometer properly.

A Model Rocket Launcher

Objective
A simple model rocket launcher and firing system can be made from household tools and a few items from a hardware store.

Materials
- Drill with a 0.125 in bit
- Wire stripper and pliers
- 12–15 feet of electrical wire, like an extension cord
- Two alligator battery clips
- Two small microclips
- Doorbell button or push button switch
- Empty coffee or similar large size can or ceramic flowerpot
- One 1 ft square piece of plywood, 0.5 in thick or greater
- One 0.125 in diameter metal rod 3 ft long

Assembly
1. Refer to the drawing as you build the system.
2. First, make the launcher by drilling a 0.125 in hole perpendicular through the center of the plywood base. Drill or punch with a nail a hole in the center of the coffee can lid. Insert the rod into the hole in the base and then slide the can over the rod. A ceramic pot will also work well as a blast deflector.
3. Take the wire and strip off enough of the insulation on the ends to attach the clips. The battery clips can be almost any size, but the igniter clips need to be small and lightweight. Flat jaw microclips are the best.
4. About 2 feet from the end of the battery clips, cut one wire and strip the ends. Connect the ends to the push button switch. Depending on the style of the switch, you may want to mount it on a small piece of wood for better handling during the launch.
5. The battery can be a 6 volt or 12 volt dry or wet cell, or a pack of four connected 1.5 volt flashlight batteries. The battery needs to be fresh or recharged to give a good current supply.

Operation

At the launch site in the field, stretch out the wire to its full length between the launch pad and the battery. Prep the rocket, and place it on the pad. Connect the microclips to the ignitor wire. Clear the area, and get ready for the final countdown. Connect the battery clips to the battery. Give the countdown, and push the button to fire the rocket. After launch, unclip one wire on the battery. Reconnect the wire only after the next rocket is on the pad. Always read and follow the instructions that come with the rocket motors, and observe the NAR safety code in Appendix B.

Model Rocket Launcher and Firing System

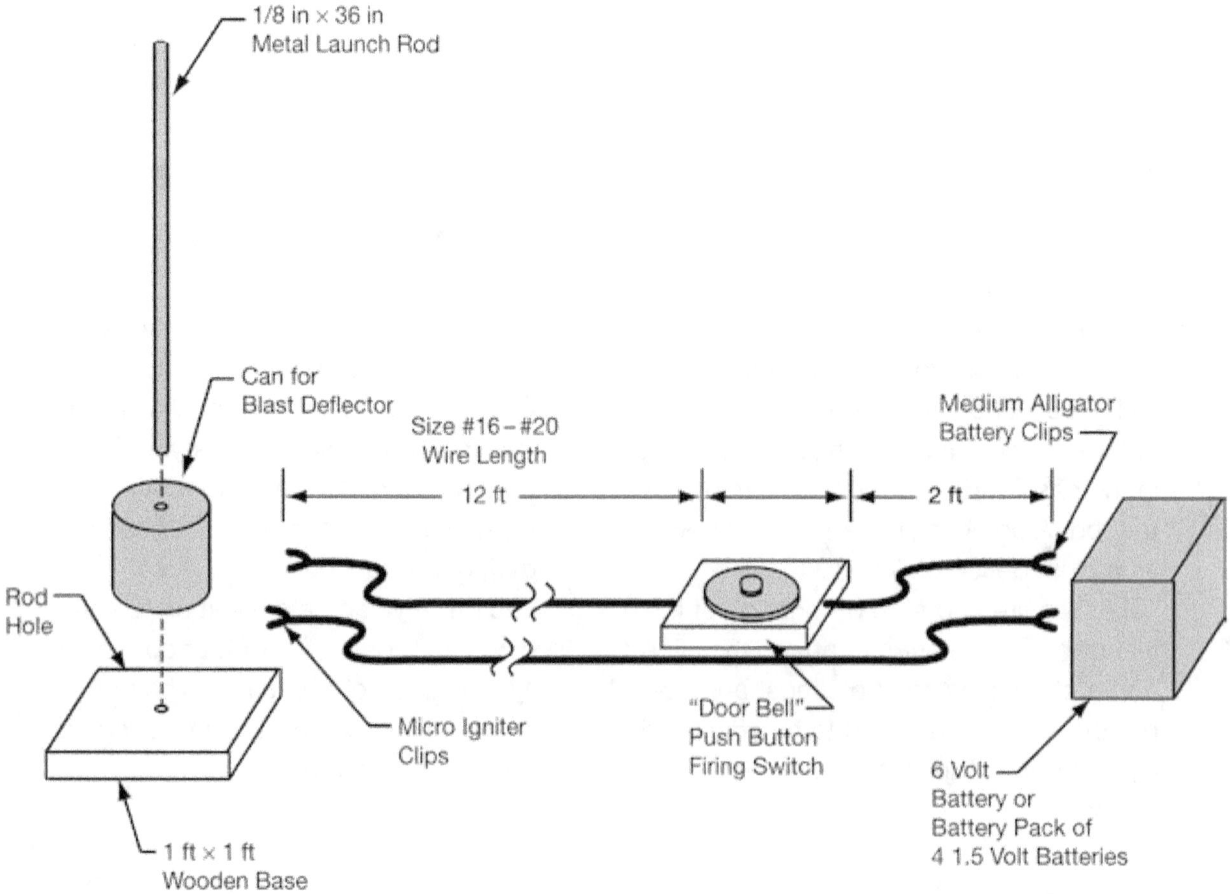

Appendix B – Rocket Safety Code

Water Rocket Safety Guide

Safety is very important with any rocket. Rockets are safe when everyone understands and abides by safe behavior. Only plastic drinking bottles should be used. New bottles should be used whenever possible. Bottles should be retired from use after 10–15 launches.

Caution: Children should be closely supervised when they are using rockets. Even if they understand and agree to the safety rules, there will be lapses in concentration or judgment. Children cannot be made responsible for the safety of others. A child may feel it is enough to tell a two-year old to stay out of the way.

Launch Safety Instructions

- Select a grassy field or athletic practice field that measures at least 100 ft (30.48 m) in width. Place the launcher in the center of the field, and anchor it in place. Caution: If it is a windy day, place the launcher closer to the side of the field from which the wind is blowing so that the rocket will drift onto the field as it descends.
- As you set up your rocket on the launch pad, observers should stand back several feet (meters). It is recommended that you rope off the launch site.
- Do not point your water rocket at another person, animal or object. Water rockets take off with a good deal of force from the air pressure and weight from the water.
- The team member responsible for pumping air into the rocket should wear eye protection. The bottle rocket should be pumped no higher than about 50 psi (3.52 kg-fscm) and never above 90 psi (6.33 kg-fscm). Before launching, consult the pressure-distance table below.
- When pressurization is complete, everyone should stand behind the roped-off area for the countdown. Two-liter bottles can weaken and will explode. Bottles should be retired from use after 10–15 launches.
- Continue to countdown and launch the rocket only when the recovery range is clear.
- If you do not experience successful liftoff, remember that the bottle is pressurized and may blast off when you touch it. Be careful; do not let it hit you. **NEVER** stand over a rocket.
- A team member should retrieve the rocket.

Table of distances for a given pressure.

Typical Classroom Maximums			
Pressure		Distance	
(psi)	(kPa)	(ft)	(m)
20	0.1379	85.30	26.00
40	0.2758	167.32	51.00
60	0.4137	252.62	77.00
80	0.5516	334.65	102.00

NAR Model Rocket Safety Code

Revised February 2001

1. **Materials.** I will use only lightweight, nonmetal parts for the nose, body, and fins of my rocket.

2. **Motors.** I will use only certified, commercially made model rocket motors, and will not tamper with these motors or use them for any purposes except those recommended by the manufacturer.

3. **Ignition System.** I will launch my rockets with an electrical launch system and electrical motor igniters. My launch system will have a safety interlock in series switch the launch switch and will use a launch switch that returns to the "off" position when released.

4. **Misfires.** If my rocket does not launch when I press the button of my electrical launch system, I will remove the launcher's safety interlock or disconnect its battery and will wait 60 seconds after the last launch before allowing anyone to approach the rocket.

5. **Launch Safety.** I will use a countdown before launch and will ensure that everyone is paying attention and is a safe distance of at least 15 ft (4.572 m) away when I launch rockets with D motors or smaller and 30 ft (9.14 m) when I launch larger rockets. If I am uncertain about the safety or stability of an untested rocket, I will check the stability before flight and will fly it only after warning spectators and clearing them away to a safe distance.

6. **Launcher.** I will launch my rocket from a launch rod, tower, or rail that is pointed to within 30 degrees of the vertical to ensure that the rocket flies nearly straight up, and I will use a blast deflector to prevent the motor's exhaust from hitting the ground. To prevent accidental eye injury, I will place launchers so that the end of the launch rod is above eye level or will cap the end of the rod when it is not in use.

7. **Size.** My model rocket will not weigh more than 53 oz (1,500 g) at liftoff and will not contain more than 4.4 oz (125 g) of propellant or 71.9 lb-s (320 N-s) of total impulse. If my model rocket weighs more than 1 lb (453 g) at liftoff or has more than 4 oz (113 g) of propellant, I will check and comply with Federal Aviation Administration regulations before flying.

8. **Flight Safety.** I will not launch my rocket at targets, into clouds, or near airplanes, and will not put any flammable or explosive payload in my rocket.

9. **Launch Site.** I will launch my rocket outdoors, in an open area at least as large as shown in the accompanying table, and in safe weather conditions with wind speed no greater than 20 mi/hr (32.19 km/hr). I will ensure that there is no dry grass close to the launch pad and that the launch site does not present risk of grass fires.

10. **Recovery System.** I will use a recovery system, such as a streamer or parachute, in my rocket so that it returns safely and

undamaged and can be flown again, and I will use only flame-resistant or fireproof recovery system wadding in my rocket.

11. **Recovery Safety.** I will not attempt to recover my rocket from power lines, tall trees, or other dangerous places.

Launch Site Dimensions

Installed Total Impulse		Equivalent Motor Type	Minimum Site Dimensions	
lb-sec	(N-sec)		(ft)	(m)
0.00–0.28	0.00–1.25	1/4A, 1/2A	50	15.24
0.28–0.56	1.26–2.5	A	100	30.48
0.56–1.12	2.51–5.00	B	200	60.96
1.12–2.25	5.01–10.00	C	400	121.92
2.25–4.49	10.01–20.00	D	500	152.40
4.50–8.99	20.01–40.00	E	1,000	302.40
8.99–17.97	40.01–80.00	F	1,000	302.40
17.98–35.95	80.01–160.00	G	1,000	302.40
35.95–71.90	160.01–320.00	Two Gs	1,500	457.20

Appendix C – Rocket Principles

A rocket in its simplest form is a chamber enclosing a gas under pressure. A small opening at one end of the chamber allows the gas to escape, and, in doing so, provides a thrust that propels the rocket in the opposite direction. A good example of this is a balloon. A balloon's rubber walls compress the air inside the balloon. The air pushes back so that the inward and outward pressing forces balance. When the nozzle is released, air escapes through it, and the balloon is propelled in the opposite direction.

When we think of rockets, we rarely think of balloons. Instead, our attention is drawn to the giant vehicles that carry satellites into orbit and spacecraft to the Moon and planets. Nevertheless, there is a strong similarity between the two. The only significant difference is the way the pressurized gas is produced. With space rockets, burning propellants, which can be solid or liquid in form or a combination of the two, produce the gas.

One of the interesting facts about the historical development of rockets is that while rockets and rocket-powered devices have been in use for more than 2,000 years, it has been only in the last 300 years that rocket experimenters have had a scientific basis for understanding how they work.

The science of rocketry began with the publishing of a book in 1687 by the great English scientist Sir Isaac Newton. His book, entitled *Philosophiae Naturalis Principia Mathematica*, described physical principles in nature. Today, Newton's work is usually just called the *Principia*.

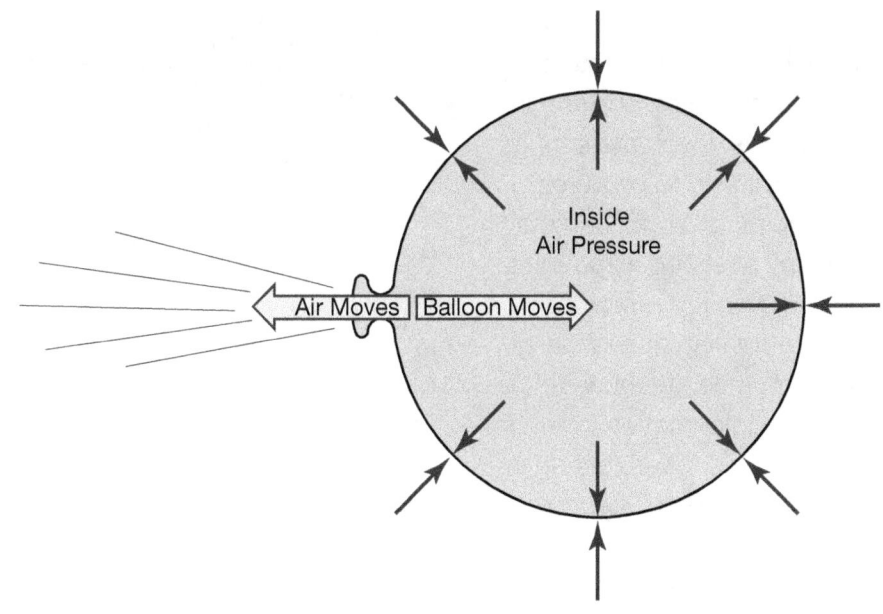

In the *Principia*, Newton stated three important scientific principles that govern the motion of all objects, whether on Earth or in space. Knowing these principles, now called Newton's Laws of Motion, rocketeers have been able to construct the modern giant rockets of the 20th century, such as the Saturn V and the Space Shuttle. Here are Newton's Laws of Motion in simple form:

1. Objects at rest will stay at rest, and objects in motion will stay in motion in a straight line, unless acted upon by an unbalanced force.
2. Force is equal to mass (m) × acceleration (a).
3. For every action, there is always an opposite and equal reaction.

As will be explained shortly, all three laws are really simple statements of how things move, but with them, precise determinations of rocket performance can be made.

Newton's First Law
This law of motion is just an obvious statement of fact, but to know what it means, it is necessary to understand the terms rest, motion, and unbalanced force.

Rest and motion can be thought of as being opposite to each other. Rest is the state of an object when it is not changing position in relation to its surroundings. If you are sitting still in a chair, you can be said to be at rest. However, this term is relative. Your chair may actually be one of many seats on a speeding airplane. The important thing to remember here is that you are not moving in relation to your immediate surroundings. If rest were defined as a total absence of motion, it would not exist in nature. Even if you were sitting in your chair at home, you would still be moving, because your chair is actually sitting on the surface of a spinning planet that is orbiting a star. The star is moving through a rotating galaxy that is, itself, moving through the universe. While sitting still, you are, in fact, traveling at a speed of hundreds of miles or kilometers per second.

Motion is also a relative term. All matter in the universe is moving all the time, but in the first law, motion here means changing position in relation to surroundings. A ball is at rest if it is sitting on the ground. The ball is in motion if it is rolling. A rolling ball changes its position in relation to its surroundings. When you are sitting in a seat in an airplane, you are at rest, but if you get up and walk down the aisle, you are in motion. A rocket blasting off the launch pad changes from a state of rest to a state of motion.

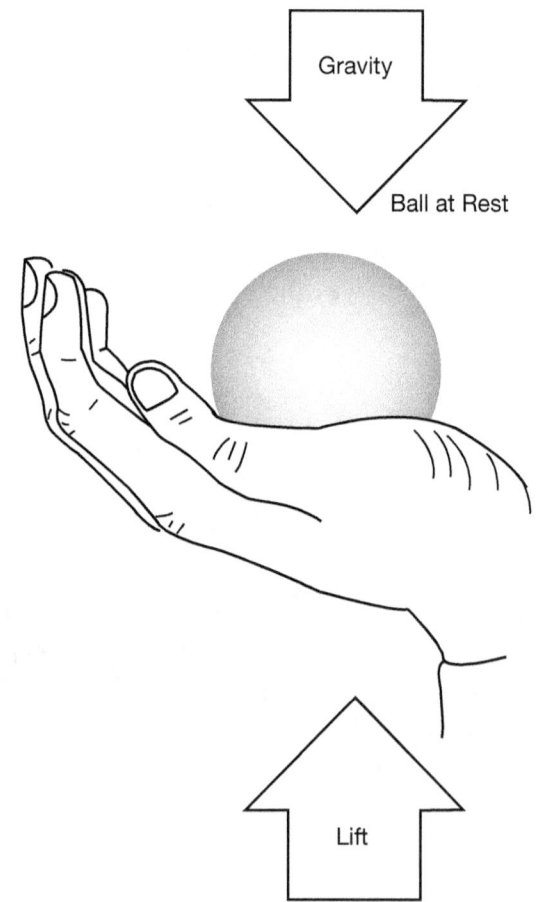

The third term important to understanding this law is unbalanced force. If you hold a ball in your hand and keep it still, the ball is at rest. However, all the time the ball is held there, forces are acting on it. The force of gravity is trying to pull the ball downward, while at the same time your hand is pushing against the ball to hold it up. The forces acting on the ball are balanced. Let the ball go or move your hand upward, and the forces become unbalanced. The ball then changes from a state of rest to a state of motion.

In rocket flight, forces become balanced and unbalanced all the time. A rocket on the launch pad is balanced. The surface of the pad pushes the rocket up, while gravity tries to pull it down. As the engines are ignited, the thrust from the rocket unbalances the forces, and the rocket travels upward. Later, when the rocket runs out of fuel, it slows down, stops at the highest point of its flight, and then falls back to Earth.

Objects in space also react to forces. A spacecraft moving through the solar system is in constant motion. The spacecraft will travel in a straight line if the forces on it are in balance. This happens only when the spacecraft is very far from any large gravity source, such as Earth or the other planets and their moons. If the spacecraft comes near a large body in space, the gravity of that body will unbalance the forces and curve the path of the spacecraft. This happens, in particular, when a satellite is sent by a rocket on a path that is tangent to the planned orbit about a planet. The unbalanced gravitational force causes the satellite's path to change to an arc. The arc is a combination of the satellite's fall inward toward the planet's center and its forward motion. When these two motions are just right, the shape of the satellite's path matches the shape of the body it is traveling around. Consequently, an orbit is produced. Since the gravitational force changes with height above a planet, each altitude has its own unique velocity that results in a circular orbit. Obviously, controlling velocity is extremely important for maintaining the circular orbit of the spacecraft. Unless another unbalanced force, such as friction with gas molecules in orbit or the firing of a rocket engine in the opposite direction, slows down the spacecraft, it will orbit the planet forever.

Now that the three major terms of this first law have been explained, it is possible to restate this law. If an object, such as a rocket, is at rest, it takes an unbalanced force to make it move. If the object is already moving, it takes an unbalanced force to stop it, change its direction from a straight-line path or alter its speed.

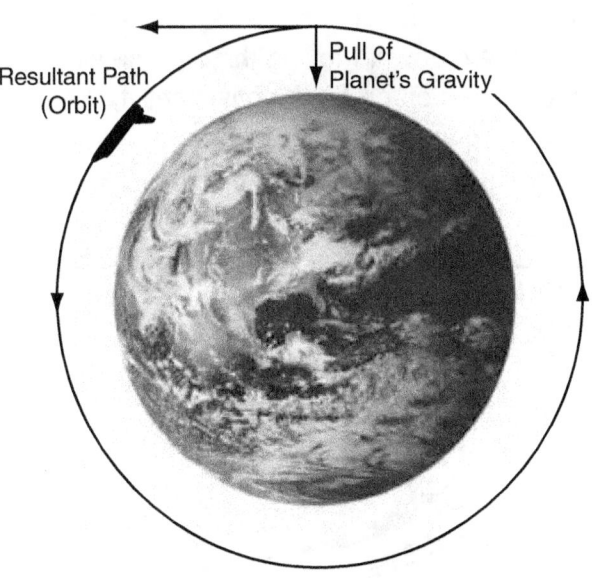

Newton's Third Law

For the time being, we will skip the Second Law and go directly to the Third. This law states that every action has an equal and opposite reaction. If you have ever stepped off a small boat that has not been properly tied to a pier, you will know exactly what this law means.

A rocket can lift off from a launch pad only when it expels gas out of its engine. The rocket pushes on the gas, and the gas in turn pushes on the rocket. The whole process is very similar to riding a skateboard. Imagine that a skateboard and rider are in a state of rest (not moving). The rider jumps off the skateboard. In the Third Law, the jumping is called an action. The skateboard responds to that action by traveling some distance in the opposite direction. The skateboard's opposite motion is called a reaction. When the distance traveled by the rider and the skateboard are compared, it would appear that the skateboard has had a much greater reaction than the action of the rider. This is not the case. The reason the skateboard has traveled farther is that it has less mass than the rider. This concept will be better explained in a discussion of the Second Law.

With rockets, the action is the expelling of gas out of the engine. The reaction is the movement of the rocket in the opposite direction. To enable a rocket to lift off from the launch pad, the action, or thrust, from the engine must be greater than the weight of the rocket. While on the pad, the weight of the rocket is balanced by the force of the ground pushing against it. Small amounts of thrust result in less force required by the ground to keep the rocket balanced. Only when the thrust is greater than the weight of the rocket does the force become unbalanced and the rocket lifts off. In space, where unbalanced force is used to maintain the orbit, even tiny thrusts will cause a change in the unbalanced force and result in the rocket changing speed or direction.

One of the most commonly asked questions about rockets is how they can work in space where there is no air for them to push against. The answer to this question comes from the Third Law. Imagine the skateboard again. On the ground, the only part that air plays in the motions of the rider and the skateboard is to slow them down. Moving through the air causes friction or, as scientists call it, drag. The surrounding air impedes the action-reaction.

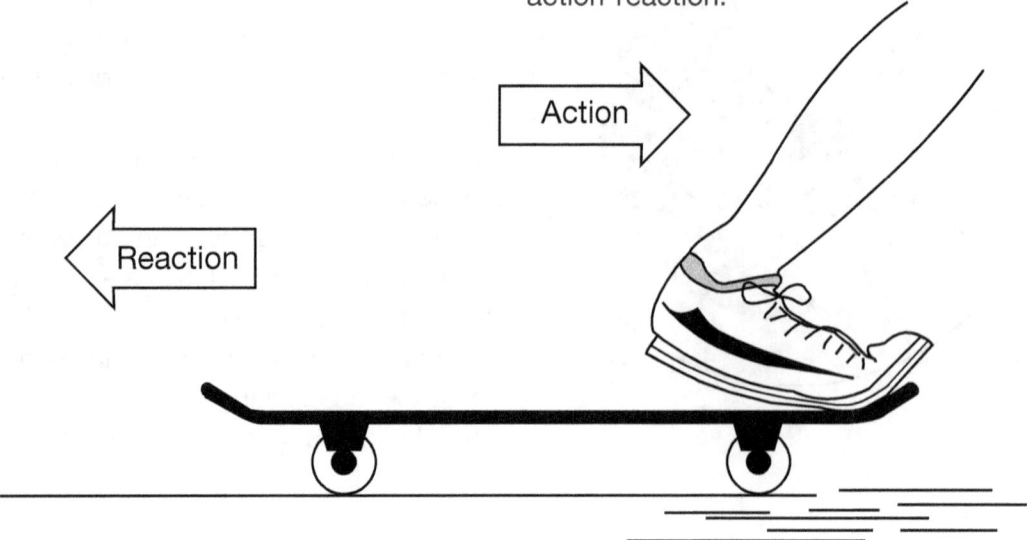

As a result, rockets actually work better in space than they do in air. As the exhaust gas leaves the rocket engine, it must push away the surrounding air; this uses up some of the energy of the rocket. In space, the exhaust gases can escape freely.

Newton's Second Law
This law of motion is essentially a statement of a mathematical equation. The three parts of the equation are mass, acceleration, and force. Using letters to symbolize each part, the equation can be written as follows:

$f = ma$

The equation reads: force equals mass times acceleration. To explain this law, we will use an old style cannon as an example.

When the cannon is fired, an explosion propels a cannon ball out the open end of the barrel. It flies several yards to its target. At the same time, the cannon itself is pushed back several yards. This is action and reaction at work (Third Law). The force acting on the cannon and the ball is the same. The Second Law determines what happens to the cannon and the ball. Look at the following two equations:

$f = m(\text{cannon})a(\text{cannon})$

and

$f = m(\text{ball})a(\text{ball})$.

The first equation refers to the cannon and the second to the cannon ball. In the first equation, the mass is the cannon itself, and the acceleration is the acceleration of the cannon. In the second equation, the mass is the cannon ball, and the acceleration is the acceleration of the cannon ball. Because the force (exploding gun powder) is the same for the two equations, the equations can be combined and rewritten as follows:

$m(\text{cannon})a(\text{cannon}) = m(\text{ball})a(\text{ball})$.

In order to keep the two sides of the equations equal, the accelerations vary with mass. In other words, the cannon has a large mass and a small acceleration. The cannon ball has a small mass and a large acceleration.

Apply this principle to a rocket. Replace the mass of the cannon ball with the mass of the gases being ejected out of the rocket engine. Replace the mass of the cannon with the mass of the rocket moving in the other direction. Force is the pressure created by the controlled explosion that takes place inside the rocket's engines. That pressure accelerates the gas in one direction and the rocket the opposite direction.

Some interesting things happen with rockets that do not happen with the cannon and ball in this example. With the cannon and cannon ball, the thrust lasts for just a moment. However, the thrust for the rocket continues as long as its engines are firing. Furthermore, the mass of the rocket changes during flight. Its mass is the sum of all its parts. A rocket's parts include engines, propellant tanks, payload, control system and propellants. By far, the largest part of the rocket's mass is its propellants, but that amount constantly changes as the engines fire. That means that the rocket's mass gets smaller during flight. In order for the left side of our equation to remain in balance with the right side, acceleration of the rocket has to increase as its mass decreases. That is why a rocket starts off moving slowly and goes faster and faster as it climbs into space.

Newton's Second Law of Motion is especially useful when designing efficient rockets. To enable a rocket to climb into low-Earth orbit, it is necessary to achieve a speed in excess of 17,398.39 mph (28,000 kmph). A speed of over 25,010.19 mph (40,250 kmph), called escape velocity, enables a rocket to leave Earth and travel out into deep space. Attaining space flight speeds requires the rocket engine to achieve the greatest action force possible in the shortest time. In other words, the engine must burn a large mass of fuel and push the resulting gas out of the engine as rapidly as possible.

Newton's Second Law of Motion can be restated in the following way: the greater the mass of rocket fuel burned and the faster the gas produced can escape the engine, the greater the thrust of the rocket.

Putting Newton's Laws of Motion Together
An unbalanced force must be exerted for a rocket to lift off from a launch pad or for a craft in space to change speed or direction (First Law). The amount of thrust (force) produced by a rocket engine will be determined by the rate at which the mass of the rocket fuel burns and the speed of the gas escaping the rocket (Second Law). The reaction, or motion, of the rocket is equal to and in the opposite direction of the action, or thrust, from the engine (Third Law).

Appendix D – Rocketry Resource Materials

Who is NAR?
Who is the National Association of Rocketry (NAR)? The NAR is the organized body of rocket hobbyists. Chartered NAR sections conduct launches, connect modelers, and support all forms of sport rocketry. NAR was founded in 1957 to help young people learn about science and math through building and safely launching their own models.

NAR Sections and its members host hundreds of launches each year—both sport launches and competitions—from the local and regional levels to national events. Designing, building and flying rockets is always more fun when you are doing it with friends.

The NAR membership consists mainly of adults, but also includes many young people and families. The most experienced model rocketeers who fly every kind of sport rocket are in NAR, and they launch together with the newbies who have just joined and discovered that they like rockets. The NAR gives its members the chance to improve and advance their rocketry skills by the association and participation in member activities and group projects.

As a national organization, NAR approves and certifies the hobby rocket motors that consumers use. NAR developed the safety code for both model and high-power sport rocketry that has protected users for 50 years. Other member benefits include NAR insurance coverage and the chance to attend NAR conferences, launches and contests. Members receive the bimonthly *Sport Rocketry* magazine that NAR produces.

How NAR Can Help You
The NAR is a nationwide network of local clubs with experienced rocketeers available to provide advice and launch sites for your flights. They welcome beginners and students who want to enjoy rocketry and learn to fly safely. Many adult NAR members are mentors and assist individuals or schools in their local area.

Please visit the NAR Web site at <www.nar.org and select http://www.nar.org/tarcmentors.pdf> for the current list of Mentors. The section directory is at <http://www.nar.org/NARseclist.php> to help you find the closest club. If you would like to have a club closer to you, NAR can show you how to start your own section.

Team America Rocketry Challenge
Aerospace Industries Association (AIA) and the NAR are proud to sponsor the annual Team America Rocketry Challenge (TARC), the largest model rocket showcase on the planet (see <www.rocketcontest.org>). The Challenge is to design, build and fly a model rocket carrying a raw egg and return it safely to the ground, while staying aloft for a specific time and reaching a specific altitude. Teams whose score is in the 100 best are invited to compete for a share of the $60,000 prize package at the National Finals at The Plains, Virginia. NAR mentors can help you form a team and learn to launch eggs in this contest.

Rocketry Resource Materials
Electronic Resources
The following listing of Internet addresses will provide users with links to educational materials throughout the World Wide Web (WWW) related to rocketry.

National Association of Rocketry
<http://www.nar.org/>

Rockets Educator Guide
EG-2003-01-108-HQ
http://www.nasa.gov/audience/foreducators/topnav/materials/listbytype/Rockets.html

The activities and lesson plans contained in this educator guide for grades K–12 emphasize hands-on science, prediction, data collection and interpretation, teamwork, and problem solving. The guide also contains background information about the history of rockets and basic rocket science.

NASA Educator Resource Centers referenced for this project:
U.S. Space and Rocket Center
NASA Educator Resource Center for
NASA MSFC
One Tranquility Base
Huntsville, AL 35807
Phone: 256-544-5812
<http://erc.msfc.nasa.gov>

NASA Glenn Research Center
Mail Stop 8-1
21000 Brookpark Road
Cleveland, OH 44135
Phone: 216-433-2017
<http://www.grc.nasa.gov/WWW/PAO/html/edteachr.htm>

Beginner's Guide to Rockets
<http://exploration.grc.nasa.gov/education/rocket/bgmr.html>

This is an excellent site with interactive features and simulations describing many rocket topics with plenty of middle to high school math and science background material.

www.ingramcontent.com/pod-product-compliance
Lightning Source LLC
Chambersburg PA
CBHW080247180526
45167CB00006B/2444